THE BODY

The Key Concepts

ISSN 1747-6550

The series aims to cover the core disciplines and the key cross-disciplinary ideas across the Humanities and Social Sciences. Each book isolates the key concepts to map out the theoretical terrain across a specific subject or idea. Designed specifically for student readers, each book in the series includes boxed case material, summary chapter bullet points, annotated guides to further reading and questions for essays and class discussion

Film: The Key Concepts
Nitzan Ben-Shaul

Globalization: The Key Concepts
Thomas Hylland Eriksen

Food: The Key Concepts
Warren Belasco

Technoculture: The Key Concepts
Debra Benita Shaw

The Body: The Key Concepts
Lisa Blackman

New Media: The Key Concepts
Nicholas Gane and David Beer

THE BODY

The Key Concepts

Lisa Blackman

Oxford • New York

English edition
First published in 2008 by
Berg
Editorial offices:
First Floor, Angel Court, 81 St Clements Street, Oxford OX4 1AW, UK
175 Fifth Avenue, New York, NY 10010, USA

Berg is the imprint of Oxford International Publishers Ltd.

Library of Congress Cataloging-in-Publication Data

Blackman, Lisa, 1965-
    The body : the key concepts / Lisa Blackman.—English ed.
    p. cm.—(The key concepts, 1747-6550)
    Includes bibliographical references and index.
    ISBN-13: 978-1-84520-589-8 (cloth)
    ISBN-10: 1-84520-589-8 (cloth)
    ISBN-13: 978-1-84520-590-4 (pbk.)
    ISBN-10: 1-84520-590-1 (pbk.)
    1. Body, Human—Social aspects. 2. Body, Human—Philosophy. 3. Body
image—Social aspects. 4. Human physiology. 5. Identity (Psychology)
I. Title.

    HM636.B55 2008
    306.4—dc22
                                                              2008022351

British Library Cataloguing-in-Publication Data

A catalogue record for this book is available from the British Library.

ISBN 978 1 84520 589 8 (Cloth)
      978 1 84520 590 4 (Paper)

Typeset by JS Typesetting Ltd, Porthcawl, Mid Glamorgan
Printed in the United Kingdom by Biddles Ltd, King's Lynn

www.bergpublishers.com

CONTENTS

Introduction: Thinking through the Body   1

The Problem of Dualism   4

The Problem of the Body as Substance   5

The Body as an 'Absent Present'   6

The Transdisciplinarity of 'Body Studies'   7

Horse–Human Relations   8

The Affective Body   9

1 Regulated and Regulating Bodies   15

Introduction   15

The Sociological Body   16

The Naturalistic Body   17

The Materialist Body   20

The Socially Constructed Body   22

The Micro and the Macro   23

The Disciplined Body   25

Agency and the Body   28

The Somatically Felt Body   29

The Sleeping Body   32

Embodiment   34

Conclusion   35

2 Communicating Bodies   37

Introduction   37

Social Influence   37

Becoming (Horse–Human)   38

Body Language   41

The Body and Performance   45

Emotional Contagion   46

Self-containment and Othering   47
Affective Transmission   49
The Civilized Body   50
The Somatically Felt Body   52
The Vitalist Body   54
The Networked Body   56
The Feeling Body   56
Conclusion   58

3   Bodies and Difference   59
Introduction   59
Bodily Markers of Respectability   61
Corporeal Capital   62
Feelings and Bodily Dispositions   64
Bodily Affectivity   65
Embodiment and Media Consumption   67
The Body and Individualization   69
The Politics of Female Bodies   71
The Corporeal Turn   72
Feminine Becoming and Internalization   74
Corporeal Feminism   76
Performativity   78
Conclusion   81

4   Lived Bodies   83
Introduction   83
The Sensient Body   84
Touch   85
Taste   88
The Mouth   89
The Mouth and Taste   91
Abjection   93
Smell   95
The Articulated Body   96
Healthism and the Body   98
Self-health   99
Narratives and Bodily Matters   102
Conclusion   103

5 The Body as Enactment   105
    Introduction   105
    Process   106
    The Body-in-Movement   107
    Bodies without Organs   109
    The Molecular Body   113
    Posthuman Bodies   116
    Companion Species   118
    Actor Network Theory   121
    The Body Multiple   124
    Socialized Biology   126
    The Body and Psychiatric Culture   127
    Conclusion: Enacted Materialities   129

Conclusion: Imagining the Future of the Body within the Academy   131
    Introduction   131
    The Affective Body   132
    Immaterial Bodies   134
    Modulation   135
    Biomediation   136
    Corporeal Thinking   137
    Conclusion   138

Questions for Essays and Classroom Discussion   139
Annotated Guide for Further Reading   143
Bibliography   147
Index   157

INTRODUCTION: THINKING THROUGH THE BODY

Just what kind of human being is that biotechnologically constructed baby – the cloned child?

Rose, *The Politics of Life Itself: Biomedicine,*
Power and Subjectivity in the 21ˢᵗ Century

Is their anything *natural* about the human body? Is this still a viable concept for organizing, examining and reflecting upon the body as an object of study within the humanities? What can be said to be distinctly *human* about the body, and how do we make such differentiations? How might motivations and methods for studying the body differ when we compare approaches within the humanities to the life, biological and psychological sciences? One of the central premises that will be explored in this book is that the 'natural body', with all the assumptions this notion brings with it, is looking rather fragile and shaky. Indeed, the Dutch anthropologist AnneMarie Mol (2002) talks about 'a' body *not* as a singular, bounded entity or substance but rather as what she terms the 'body multiple'. She argues that the body is not bounded by the skin, where we understand the skin to be a kind of container for the self, but rather our bodies always extend and connect to other bodies, human and non-human, to practices, techniques, technologies and objects which produce different kinds of bodies and different ways, arguably, of enacting what it means to be human. The idea of the body as simply something that we both *have* and *are* is displaced in this perspective as the focus shifts to what bodies can *do*, what bodies could *become*, what practices enable and coordinate the *doing* of particular kinds of bodies, and what this makes possible in terms of our approach to questions about life, humanness, culture, power, technology and subjectivity. These are some of the themes we will explore throughout the book and which radically refigure the idea of the body as *substance* or *entity* and even as distinctly *human*.

Current reformulations of the body are situated within a historical conjuncture which is often characterized by its apparent *newness*. We are confronted with new ways of making and re-making bodies. The singular, bounded, carbon-based body is being replaced by the proliferation and emergence of technologies and practices which enable the enhancement, alteration and invention of new bodies. Some appear mundane and have become a largely accepted way of living and acting upon our selves and others. These include practices of body modification and enhancement such as cosmetic surgery, gender reassignment surgery and in vitro fertilization. The more far reaching are biotechnological practices which are challenging the concept of what we understand *life*, the *natural* and *humanness* to mean. These include what Nikolas Rose (2007) has termed 'technologies of enhancement and technologies of susceptibility', which are future-oriented technologies that allow intervention at the molecular level of life (codes, enzyme activities, neurotransmitters and transporter genes, for example). Their interventions, it is argued, have the potential to improve selfhood, vitality, health and lifestyle. These technologies, the mundane, speculative and challenging are not without controversy. The technique of oocyte cryopreservation has challenged conceptions of fertility and womanhood, by creating the possibility for post-menopausal women to give birth to genetically related offspring well into their fifties or sixties (Watkins 2007).

How do these reformulations of what we understand bodies to *be* and *do* reinvent questions and concepts that have been central to sociological, anthropological, psychological and cultural theorizing? These include the question of how power operates in relation to personhood, what the relationships are between technologies and identities, the extent to which bodies can be said to be material, social, ideological or cultural, for example, and how we might understand these terms. 'Thinking through the body' creates an important challenge for reimagining possible solutions to some of the frameworks which have organized theorizing across the humanities. These can be characterized as how to 'think' the relationship between the *micro* and the *macro*, the *individual* and the *social*, *structure* and *agency*, *mind* and *body* and the *inside* and *outside*, for example. The reformulations of 'the body' and bodies across the humanities have also demanded an imaginative reengagement with method. If the body is not simply a *natural* body, the rightful province of the life and biological sciences, then how can bodies be examined and interrogated through frameworks that have been understood as more social or cultural? What does it mean to offer a *social* or *cultural* analysis of bodies, and is the addition of 'the social' part of the problem in 'thinking through the body'? All of these questions and more were taken up in the emerging 'sociology of the body' in the 1980s, which forms the basis of Chapter 1. This is not where 'body theory' begins, as, arguably, the question of female bodies, their basis, status and potentiality has always been central to feminist theorizing. However, sociology of the body has become an accepted tradition within sociological studies; it has formed the basis of new methods and concepts for examining the corporeality of the social and the sociality of corporeality. It also provides a link to questions, methods and concepts mobilized across cultural theory, critical psychology, science studies, anthropology and related perspectives.

As we will see, the question of the place of the body within social theorizing relates to a much broader question about exactly what we mean when we talk about bodies, or call for the development of 'body studies' as a central line of questioning and reflection. Thus, to 'think through the body' also demands critical reflection and examination in relation to this key question. This book will engage with that central question, and the call from many disciplines to re-embody theory by exploring how different perspectives allow us to approach this question in countless different ways. As we will see throughout the book, there is not a coherence that unites the different perspectives, but most, if not, all start from the premise that 'to achieve an adequate analysis of the body we need to regard it as a material, physical and biological phenomenon which is irreducible to immediate social processes or classifications' (Shilling 1993: 10). This gesture towards the interplay of biological, physical and social processes may not seem so radical in the context of how we might embody our own sense of subjectivity. For example, we may feel that of course our sense of who we are is an amalgam of our physicality, biological processes and our place and position as particular kinds of social subject. However, we will see that the question of how to bring the biological and physical together with the social is not merely linked to the recognition of this intersection. One of the problems central to the call to 'think through the body' is exactly how we can bring together processes that are often viewed as separate entities. Indeed this assumption of separation is one that historically has underpinned the development of the kinds of disciplinary specialization that have led to a split between the natural and human sciences.

Sociology, for example, was framed by some of the early founding figures of the discipline as an examination of *social* reproduction: of how ideas, beliefs, practices, traditions and so forth are reproduced in such a way that they appear uniform and become part of the social fabric. Sociology took as its object what was considered *social* about the ties and obligations that bind individuals and groups to each other. This assumed a separation between the biological and the social that was reproduced in the differentiation between the natural and human sciences, with the former focusing upon what was taken to be distinctly biological about what it means to be human. As we will see, this question of what disciplines can rightfully claim as their object and subject matter is central to exactly what we might mean within the humanities when we call for the body to be taken seriously. One of the focuses of this book will be on the various ways in which humanities scholars have attempted to bridge this schism and produce work that starts from the assumption that, if we are to work towards adequately re-embodying social and cultural theorizing, this split and separation is part of the challenge.

THE PROBLEM OF DUALISM

One of the key splits or separations that have been reproduced in different ways across the natural and human sciences is the mind–body dualism. We will encounter this dualism and its revision and rethinking in different ways throughout the chapters that make up this book. However, let us start by considering what is meant by the concept of dualism, and how it might elaborate the relationship between the mind and body. The mind is often used to refer to and make possible those processes that allow us to think, reason, argue, reflect, debate and write. These processes of thinking are usually referred to as *cognitive* – those that arise from the mind's ability to engage in activities that are distinct from those located within the body, such as respiration and digestion. The latter processes are largely viewed as involuntary. This mind–body dualism is also often known as *Cartesian dualism* with reference to the writings of the seventeenth-century philosopher, René Descartes. Descartes argued that rationality was the key determinant of human existence. He termed this the *cogito* which is enshrined in the saying 'I think therefore I am.' This foundational dualism has a number of other key dualisms which overlay it. For example, we can already see how another dualism maps onto the separation of mind and body; the idea that the mind is subject to voluntary control, usually characterized as *will*, and the body is subject to laws which govern and regulate processes which do not require conscious effort or attention. This distinction between what is taken to be involuntary (and therefore fixed), and what is taken to be voluntary (and therefore subject to change) produces the mind and body as distinct entities. The mind is the location of thought and the body the location of a fixed set of physiological processes. But of course that is not all there is to say.

One example that complicates this distinction is revealed in a newspaper article from the early 2000s which discusses what we might refer to as the 'Aha' experience, which was identified by the author of the article and possibly shared by some of you yourselves. Perhaps after expending a lot of time trying to work through a problem, you may have given up and fallen asleep, only to awake with a solution. The author refers to this manner of problem-solving as a magical Aha experience, and relates it to a mysterious place which is located within the unconscious. Although cognition is largely viewed as voluntary, this example introduces a reformulation of thought as being both conscious and unconscious. The boundaries between what is voluntary and involuntary now start to appear less certain, more shaky and more difficult to differentiate and distinguish. This example, which could be extended to other very commonplace experiences, such as feeling moved by a film but finding it difficult to say why, capture some of the tensions which studies of the body are making visible, in particular the idea that the body is the container of a fixed set of physical processes, and that the mind is more fluid and subject to cultural influence. Although this recognizes how the mind, at least, is not fixed and exists within a cultural milieu that shapes cognition, the body is closed off from cultural analysis and seems to have nothing to offer to the disciplines of sociology, cultural studies, critical psychology and allied perspectives.

THE PROBLEM OF THE BODY AS SUBSTANCE

We can also interrogate this dualism further in the context of academic study. The exercise of rationality is often aligned with those practices linked to academic study, where the academic project is often viewed as a work of thought. This presumes that thinking primarily takes place independently of the body. I am sitting here writing this introduction, you are reading it, perhaps hoping that you will be able to make better sense of what can be a confusing field of study. We all have bodies; this is the commonsense response to what is often seen as the relegation of the body to the work of thought. Whilst writing this I am aware of my posture, of how I embody the movement of my fingers on the keyboard through my musculature and my skeleton. You might be aware of your digestion, your respiration, your nervous system, or at least your focus of attention might now have shifted to these processes which continue, often beyond your conscious awareness. Is this the body which we wish to include in social and cultural theory? The answer is neither a simple 'yes' or 'no'. It is rather more complicated, and in order to address what 'thinking through the body' might mean we need to be aware both of the bodily basis of thought and the cognitive component of bodily processes and vice versa. We also need to move beyond thinking of bodies as substances, as special kinds of *thing* or entities, to explore bodies as sites of potentiality, process and practice.

We also need to be mindful of what happens to our views of bodies if we continue with the tradition that would locate reflection and reason within the mind. What happens to the body within these formulations? What do we make of the view that many great philosophers who located thought within the mind, as distinct from the body, suffered from bodily infirmities and diseases which helped to produce an omnipotent fantasy that the body could be overcome or even dispensed with? This is embodied in the maxim, 'mind over matter' (Porter 2003). But the body does matter and as we will see throughout this book, it is taken to matter in countless different ways. One way the body matters that I have already alluded to is in the writing and reading process itself. When we write or read we take up particular bodily orientations; posture, musculature, breathing, and certain habits or dispositions. We do not simply think, but relate to the keyboard or book through particular bodily dispositions and practices. These might appear to be automatic or involuntary, but, nevertheless, the body is never simply left behind within academic study. Indeed, it is made to relate to itself and others in particular ways through the manner in which it is situated within space (the library or lecture theatre for example), or time (where the body's cyclical rhythms for sleep, food and so forth may be ignored and overridden).

THE BODY AS AN 'ABSENT PRESENT'

Though it provides an incomplete picture, the view I have just reproduced of the body is one that entirely fits within the dualism or separation of the mind or body. Although I might be recognizing the contribution of the body to the process of thought, the body is itself assumed to be almost machine-like in its formulation. That is, that the body is a mere physical substance that, although a silent presence in thought, can, through an act of will or recognition, be attended to so that it is not taken for granted. We can all become more aware of that which is often distorted or forgotten as we go about our daily lives. This is yet another variant of the mind over matter argument that always elevates the mind and thinking to that which is superior, in relation to a conception of the body that tends towards its formulation as a physico-chemical adjunct to thought. This dualism is insidious and rather difficult to think against, and it appears in different ways across the humanities. One way that should be familiar to most readers of this book is in work of transdisciplinary relevance that has taken as its focus the role of cultural symbols and codes in the formation of identities. This work has often come out of semiotic, or what are often referred to as social constructionist, traditions (see Chapter 1). Within these traditions, if we want to analyse the role of social and cultural processes in our formation, then the obvious route is to explore the different interpretations that we might give in response to particular events in our lives. This is the site of culture as it intersects with our sense-making activity. Our bodies are there, for sure, they may register our anger, our surprise, our joy, our hurt, our pain or our suffering. However, they are merely containers for experiences, which are a product of the ways in which we use particular cultural narratives and interpretations to make sense of our lives. Culture is about sense-making. Although the *sense* in *sense*-making might make us think of a more sensient *body* it is generally linked to interpretation, to judgement and ultimately to the work of thought. We are back with culture from the neck up, as a famous scholar once said, and the body seems to have disappeared again, or at least to merely be an *absent presence*.

THE TRANSDISCIPLINARITY OF 'BODY STUDIES'

So far in this introduction you have been introduced to the importance of the concept of dualism for 'thinking through the body'. I have tried to give fairly commonplace examples that most of you will be familiar with, and to begin the work of unsettling many of the presumptions that you will live and embody in your own lives. For talk of the body is also inevitably a foray into some of the assumptions that we all make about bodies; those of ourselves and others, human and non-human, and thus can never be simply an intellectual exercise. Hopefully, the very idea of something itself being an intellectual and therefore non-bodily exercise should, even on a fairly banal level, be more difficult to entertain. However, there are certain concepts that it is important to grasp in order that you can find your way through the vast literature that is now beginning to be referred to as 'body theory' or the area of 'body-studies' (Shilling 2003). This will include some of the concepts that are used by various scholars to talk about the body; these include the *corporeal*, the *somatic* and the *material*. We will encounter these concepts and their value in revising what we mean by the body in Chapter 1. The concepts are often viewed as interchangeable, but in some instances have a very specific meaning and purchase. Again, these conceptions are always situated differently and travel in different ways across different paradigms and positions. The literature taken to comprise the growing area of 'body studies' is markedly *transdisciplinary* and crosses over the borders and boundaries between psychology, sociology, cultural theory, anthropology and sociology. You may already, therefore, have a training in particular concepts, and may or may not have already considered their value in producing accounts of what we might mean when we call for the body to be taken seriously within the humanities.

This book will not assume a particular intellectual trajectory or canon, but rather will attempt to draw out the key concepts for navigating the literature. The concepts we will make visible and use to think through some of the problems that the body presents for social and cultural theory will be found in obvious and less obvious places. Mystery, wonder and intrigue will undoubtedly mark this journey. In order to ease you into some of the exciting paradoxes and problems that studies of the body are making visible, I want to continue these introductory remarks by considering further the problem of dualism. So far, we have encountered this problem in relation to the mind and body, with a number of other key dualisms making an appearance. These include the voluntary and involuntary, the natural and the cultural and the individual and the social. The very concept of dualism, as you will have already seen, relies upon a notion of separation. Separation assumes clearly bounded entities; the biological and the social, for example. These are taken to exist and somehow interact or come together in a rather peripheral fashion. Separation assumes that it is relatively easy, if you have the expertise or knowledge, to designate what is biological and what is social, what is human and what is non-human, what is voluntary and what is involuntary. Let us follow the logic of this argument further and consider an examination by an Australian sociologist of the problem of separation in relation to horse–human relationships. This problem is confounded by the example she recounts of what enabled her horse to recover from a paralysis that prevented her from being able to ride her.

HORSE–HUMAN RELATIONS

> Connectings between human and animal are creative processes of coming to be. Putting into question humanist assumptions, I propose that we are always already part horse, and horses, part human; there is no such thing as pure horse or pure human. The human body is not simply human.
>
> Game, 'Riding: Embodying the Centaur'

The account that Game gives of her experience of helping her horse to regain its strength and ability to canter, trot and be able to take the weight of a rider was published in a British journal, *Body and Society*. This journal is one of the most respected academic journals for publishing transdisciplinary work on the body. I say this because many of you may find the idea of the human body not simply being human rather strange, perhaps disconcerting, or even amusing. However, the personal story that Game recounts of her experiences reveals how central the idea of separation is to how we tend to think of our own body and those of others, human and non-human. She begins her argument by drawing attention to how, in popular discourse, and indeed in many popular science books, such as the book *Dogs That Know When Their Owners are Coming Home* (Sheldrake 2000), we are used to the idea of some kind of mixing or intimate connection between pets and their owners. Many pet owners, and indeed you may have had this experience yourself, attest to the subtle capacity of pets to be able to connect with the moods and emotions of their owners. Some even describe their pets as psychic, as having some kind of special sense that enables communication to take place even when they are physically separated. Indeed, this is the argument that Rupert Sheldrake makes in his account of the apparent psychic abilities of animals in relation to humans. Game suggests that this idea of mixing between the human and the animal has a long history and is embodied in the mythical creature known as the centaur. The centaur was half human and half horse. So, actually, this idea of interconnection or mixing is not new or even that strange, but it does, as we will see, challenge the idea of separation that is so integral to the Western, individualized self.

Game describes how her eventual successful attempt, after much hard work and struggle, to help her horse, KP, to learn how to move again, and then to trot, canter and take the weight of a rider, came about through *forgetting* that she was separate from the animal. She equates this forgetting to a letting go of self-consciousness. This involved what she describes as a very 'relaxed concentration, a very focused and meditative state' (Game 2001: 8). In this state she was able to mount KP, and to try and connect with subtle movements that the horse was making in order to help her to remember what it felt like to canter and trot with a rider. This intimate connection is described by Game as a form of *attunement* or *entraining*. Both of these concepts are derived from more spiritual traditions such as Buddhism, for example, and assume that it is possible to develop a kind of 'sensitive feel' where you can connect in very subtle ways with those around you. Indeed, the notion of a sensitive feel was taken up by R. D. Laing (1985), the famous British anti-psychiatrist who first contended in the 1960s that when we are open to the presence of somebody we often start breathing in synchrony. What is important is not separateness, but *rhythm* and the flow of rhythms from those you are in connection with; human and non-human. This is a form of 'tuning in' that is felt in the body, and when in synchrony may be described as just 'feeling right', of being in tune with somebody, for example. In order to help KP, then, Game entrained with the rhythm of the horse, with the result that she could describe the process as 'imagining the rhythm, feeling it in our bodies, taking it up in relation with the horse, riding into the rhythm' (2001: 8). This was certainly not will, the idea of mind over matter that we have already encountered, but a letting go of the very idea that the horse and human were separate entities. As she says, 'To help her to remember canter, my body had to take up this movement. The between horse-and-human movement of canter had to be generated for KP to entrain with it, to get in the flow' (Game 2001: 6).

THE AFFECTIVE BODY

This introduces a more affective element to 'thinking through the body' that we will explore in more depth in the chapters to come. But certainly, the kind of body that Game suggests is important for sociology and cultural theory is one that is not simply inert mass. It has vitality, an aliveness that provides the potential to connect in ways that trouble and challenge the very mind–body dualism that we have already encountered. This body is what Sheets-Johnston refers to as the 'somatically felt body' (1992: 3), and it is one that will appear in different ways throughout the book. This *felt* body is one that is never singular and never bounded so that we clearly know where we end and another begins. This is a *feeling body* that presents a challenge to the kind of Cartesian dualism that produces the body as mere physical substance. The affective body is considered permeable to the 'outside' so that the very distinction between the inside and the outside as fixed and absolute is put into question. One of the central problems is how we might think this permeability, and what this might suggest about the kind of body or bodies that have been, and are being, brought into social and cultural theorizing. I hope that this has given you some of the flavour of the challenges ahead, and indeed the anomalies that Cartesian dualism covers over and silences. These themes and issues will be taken up in different ways throughout the chapters to come, where I present some of the key concepts that organize this field of study and debate. The book is not exhaustive but considers the body in relation to some of the key features and debates that are central to the field. There are, of course, many other areas that could have been included in the book and which deserve attention. Some of these important areas that have been omitted or only covered very briefly include the body and virtual technologies, disembodiment, death, queer bodies and performativity, racialization and bodily matters, and social differences such as ageing. Those that appear more substantively in the book cover the key areas that have historically contributed to the growing area of body studies. The chapters also point towards the importance of those concepts such as affect and enactment that are likely to be central to new conceptualizations of embodiment and disembodiment. I have also chosen areas of study that travel across the disciplines of sociology, anthropology, psychology, cultural studies and science studies, and this reflects my own position as a scholar situated at the intersection of critical psychology, cultural and science studies.

REGULATED AND REGULATING BODIES

These concepts are elaborated through an engagement with the classical and contemporary sociological work on the body. The sociology of the body laid out a particular problematic and set of concepts for thinking through the body. This work is often linked to a number of key texts within the field, published in the 1980s and 1990s, which began to address the absence of the body in social theory (Featherstone, Hepworth and Turner 1991; Turner 1984, 1996; Shilling 1993, 2003; Burkitt 1999). It linked a growing awareness of the significance of the body for social theory in relation to key events taking place outside of the academy. These included the emergence of a 'risk' society (Petersen and Bunton 1997), the rise of HIV and Aids (Kinsman 1996) and the growth of an array of practices and technologies which took the body as a key site for transformation and change. These include the emergence of body practices aligned with *hard* technologies – those that directly make and remake the materiality of the body (cosmetic surgery and bio-technologies) – alongside the growing popularity of *soft* technologies – those that shape and frame different aspects of the body as key sites for identity formation and expression. Consumer culture and its growth was considered a key site for the growth and proliferation of these body practices, such as tattooing and piercing, as markers of taste and style (Featherstone *et al.* 1991).

COMMUNICATING BODIES

The concept of *social influence* – the language of two separate yet proximate domains of influence – is thoroughly entrenched in our everyday understandings of the body. This is particularly so in our lexicon of emotions and body language. These examples will be developed by considering an early experimental psychological experiment which confounded psychologists as it suggested that Hans the horse was capable of being affected by the body language of the experimenter. The reader will be introduced to the different interpretations of this experiment which have been made over the years, and a reformulation of this in recent work on the body by Vinciane Despret (2004a and b). The examples discussed will show the salience of a particular set of contrasts for thinking through the notion of body communications. These include contrasts between the authentic and the performed, the primitive and the civilized, the honest and the deceptive and the human and the non-human. These contrasts will be mapped by exploring the links between three domains: academic texts mainly derived from the discipline of psychology; the popularization of these ideas in the staging of reality T.V. shows such as *Big Brother*; and the proliferation of particular understandings of how the body communicates within the field of government, business and the media more generally.

BODIES AND DIFFERENCE

The body is related to identity in complex ways which often involve the reproduction of norms. Some of the work which engages with bodily differences is derived from the more sociologically inflected work that we will have explored in Chapter 1. Other of it has been produced by feminist inspired work across sociology and cultural studies. The chapter will review this work by exploring the particular classed

(Skeggs 1997, 2004), gendered (Butler 1990, 1993) and sexualized (Creed 1993; Butler 1990, 1993, 2004; Prosser 1998; Halberstam 1998, 2005) dimensions of techniques of bodily adornment, comportment, transformation and reproduction. This work will be considered in the context of how bodies are materialized and normalized through particular conceptions of normality and abnormality. This focus on the normative body, and the formation of identity, will draw out the broader processes of power, ideology, marginalization and inequality, which frame relationships of bodies and identities. This work troubles the notion of the *natural body*, and instead moves the study of the body into a complex array of entangled material, social and historical forces.

LIVED BODIES

This chapter will develop a number of themed studies in order to assess the contribution of work that focuses upon lived experience to studies of the body within social theory. The first theme will explore the senses and the sentient body in terms of bodies living and moving as experiencing subjects. The senses will be considered as processes that connect the body to its 'outside' rather than as fixed biological entities. The emphasis is on movement and process rather than location and interiority (Manning 2007). The second theme will be developed in the context of health and illness, and includes Jackie Stacey's study of her own cancer diagnosis (1997), and work that has taken the mouth and teeth as its object (Falk 1994; Nettleton 1992). These studies develop the concept of *abjection* for considering the intense feelings of disgust that we might have when certain borders and boundaries are crossed. These include the inside and the outside, and surface and depth. The theme of health and illness is then further considered in relation to the concepts of healthism and 'somatic individuality', in which our understandings of the body are increasingly aligned to biological and biomedical explanations. We will consider here how bodies are never fixed by a biomedical gaze and are, importantly, often lived through *narratives*. The chapter will show how a focus upon the *lived body* troubles the idea that the biological and the cultural are two separate, discrete entities. Throughout the book so far we will have been considering the broader question at the heart of studies of the body and embodiment: that is, exactly 'what is the body?' (Shilling 1993: 6).

THE BODY AS ENACTMENT

Chapter 5 will continue this trend by focusing upon work that asks not what a body is but what a body can do. This work suggests that we are never a singular body, but are multiple bodies that are brought into being and held together through complex practices of self-production. The concept of multiple bodies will be developed by exploring the work of the Dutch sociologist AnneMarie Mol (2002) and some of her collaborations with the British sociologist John Law (Mol and Law 2004). This focus upon practice, enactment and performance will be related to a study of the voice hearing phenomenon that explores how what are taken as fixed biological symptoms of a disease process can be transformed through engaging in different techniques of self-production (Blackman 2001). As we will see, these techniques are not simply cognitive and introduce the importance of affectivity for thinking about the workings of power, ideology and social processes.

THE TURN TO AFFECT

The conclusion of the book, 'Imagining the Future of the Body within the Academy' will develop this theme, we will consider the turn within social and cultural theory to affect and an affective body taking place at the time of writing. Although we will have encountered the significance of affect in different ways throughout the book, we will consider how a revised focus upon affect is being developed to explain or account for some of the problems and anomalies that 'body theory' has not been able to resolve. What we will see is that the changing status of the body is gathering momentum with the result that the earlier work with which we began the book is mutating in new and exciting directions. The theme of this new turn echoes the commitment of the earlier sociological work on the body that framed the body as always in 'a process of becoming' (Shilling 1993: 5). It connects up some of the gaps and anomalies we will have encountered throughout the book and attempts to further inject an aliveness and vitality into a body that for many years has only been an absent presence in social and cultural theory.

1 REGULATED AND REGULATING BODIES

In writing about sociology's neglect of the body, it may be more exact to refer to this negligence as submergence rather than absence, since the body in sociological theory has had a furtive, secret history rather than no history at all.

Turner, *The Body and Society: Explorations in Social Theory*

INTRODUCTION

This chapter will review some of the body concepts that were introduced within sociology in the call to take the body more seriously as an object of analysis. As we will see, the invitation to sociologists to accord the body a more central place did not mean that the body until that point had not been considered. Rather, one of the strategies of this work was to reveal to sociologists how the body had always been central to sociological analysis, albeit in a silenced and unacknowledged way. One of the key concepts it introduced, that drew attention to the central yet marginalized role of the body in theoretical work, was the notion that the body was an *absent presence*. That is, that assumptions about the contribution of 'the body' to the question of how social processes worked was implicit in the theories put forward by those considered to be the founding figures of sociology. One trend of this work is a reengagement with the concepts introduced by some of those founding figures. These are names considered to be part of the canon or intellectual heritage of contemporary sociology and they will appear in most discussions of the historical development of the discipline. They are scholars who are considered important to how the project of sociology was shaped and framed so that it became distinct from other disciplines, such as psychology or media studies, for example. Let us begin the work of this chapter by breaking down the concept of the body as an *absent presence* further, and situating it within some of the broader moves and debates that were beginning to characterize the emergence of body studies within sociology and social theory.

THE SOCIOLOGICAL BODY

The emergence of sociology has been linked to two key questions: how to account for social change, and how to account for social reproduction. Early sociology was concerned more with the latter question and is often aligned to the work of the French sociologist Emile Durkheim. Durkheim argued that sociology should be an examination of the constraint and imposition of social structures on the formation of human subjects. What puzzled sociologists, psychologists, biologists, economists and other scholars in the nineteenth century was how ideas, beliefs, practices, traditions and even emotions could spread throughout populations to such a degree that they would achieve a uniformity or social unity. Durkheim argued that sociology should be an examination of this social unity, and made the concepts of *imposition* and *constraint* central to his project. He fiercely rejected the contribution of other disciplines such as anthropology and psychology to this question, and instead reified the importance of understanding the role of social structures in the formation of what it means to be human. Durkheim's approach is now regarded as *functionalist* as social structures were seen to constrain the individuality of subjects so that there appears to be no room for manoeuvre. The term 'functionalism' identifies how the privileging of the role of social structures ignored or devalued the agency of the human subject, who was seen to be at the mercy of social institutions and practices. So what kind of body was Durkheim implicitly mobilizing within his analyses? As we can see, Durkheim starts from the presumption that institutions and the state are the seat or centre of power, that individuals are mere pawns within broader ideological processes. However, these processes are effective for Durkheim because they change or transform human subjects. The changes are realized both in the bodies and minds of subjects with the result that they become particular kinds of subject or citizen.

This approach to the formation of human subjects is also characterized as part of a tradition within sociology that focuses upon *cultural inscription*. As this term suggests, what is important is how social or cultural processes inscribe or speak through individuals. These processes are manifested in the thoughts, actions, bodily dispositions and habits of subjects with the result that they appear natural and automatic. The body, within these accounts, is important for understanding the workings of ideology and power, for example, but what is brought into the analysis, albeit in an under-theorized way, is a view of the body as a malleable entity that cannot speak back. Thus, as Turner (1991: 5) argues, forming the backdrop to sociology is an assumption that 'the body is the central metaphor of political and social order'. However, in his earlier seminal work on the subject, Turner also argues that although a central metaphor, the body tends to make a 'cryptic appearance' (1984: 2). Thus, Durkheim assumes 'the coercive nature of moral facts' (Turner 1984: 21) to be such that power is taken to work through constraint and repression of bodily and psychic processes. Durkheim was not interested in what he termed the organismic basis of the body, for bodies were always 'made social' and existed within a network of ties, obligations and duties. These social ties were what mattered, relegating 'the body' (as a possible collection of instincts, drives, desires, passions, physiological processes and so forth) to the sidelines. Although he made the argument that sociology should maintain itself as separate and distinct from work developing across the psychological and biological sciences, he did, however, increasingly turn to these areas in his later writings to reflect upon the dualism of so-called human nature (see Durkheim 1960). We can see, then, that even early sociologists found the separation between mind and body, nature and nurture, and individual and society difficult to maintain.

If we revisit the basis of some of these arguments, as other sociologists of the body have already done (Shilling 1993; Turner 1984; Featherstone *et al.* 1991), what might this tell us about the status of the body within this work? We must be careful not simply to dismiss this work as devaluing the agency of the body. As Shilling (1993) argues, one of the important insights of this work is that the body is always a body that is an *unfinished* entity. In other words, the body is not simply a body defined by a fixed human nature, but, rather, bodies can, will and do change and transform given the particular set of historical circumstances within which they are socialized. Thus, talk of the body is always talk of the social context, social practices and ideological processes that produce bodily matters. However, somewhat ironically, the body that is 'a hidden base, under-theorized and taken for granted' (Shilling 1993: 20) is also deemed to be a body that cannot be explained by understandings of its biological or physiological processes. Thus, what characterizes, and has characterized, models of cultural inscription is a distance from both biology and psychology and a reification of social structure in the importance of understanding bodily matters. This is why Turner characterizes the development of sociology as 'a somewhat hostile reaction to Darwinistic evolutionism, eugenics or biologism' (1991: 7).

THE NATURALISTIC BODY
DARWINISTIC EVOLUTIONISM

This critique of the foundations of sociology is really important for some of the moves that have been advocated in contemporary work on the sociology of the body, so we will examine it in more detail. To start with, what does it mean to dismiss the foundational assumptions of sociology as being a reaction to Darwinistic evolutionism, eugenics or biologism? We will start with the tradition of Darwinistic evolutionism and examine some of the assumptions it is taken to make about what makes us human. Darwin (1859) was also interested in the question of transmission or reproduction characterized in the process of natural selection. Natural selection referred to the processes through which physical and mental traits are passed on, become modified or disappear when viewed across generations. As well as an account of physical adaptation across time, Darwin also provided an account of already existing social hierarchies at the time of his writing in the late nineteenth century. This is by far the more controversial aspect of his writings, and one that has led to the most hostile reaction by sociologists and others, such as feminist and postcolonial writers, wanting to distance themselves from the political ramifications of his theories.

EUGENICS

Eugenics was a nineteenth-century governmental strategy that incorporated the knowledge of Darwinistic evolutionism as a way of managing and governing key social issues of the time. These included the problems of vice, criminality, madness and unemployment, for example. These problems were framed within the strategies of eugenics as a problem of *degeneracy*. Degeneracy is a term that is derived from Darwinistic evolutionism and understands certain problems as reversions to more primitive forms of behaviour and experience. Thus, the problem of madness,

---

Case Study

The idea of madness and its link to degeneracy became a coherent explanation of madness from the mid nineteenth to the early twentieth centuries. This developed a view from physiognomy that saw madness as literally 'written on the body'. Madness was taken to be expressed physically as a form of biological decay, deterioration and reversion to what were viewed as more 'primitive' modes of existence. Thus the German psychiatrist Emile Kraepelin, often heralded as developing the concept of 'dementia praecox', made the following statement about the identifying features of this disease process: 'all sorts of physical abnormalities exist with striking frequency, especially weakliness, small stature, youthful appearance, malformation of the cranium, and of the ears, high and narrow palate, persistence of the intermaxillary bone, abnormal growth of hair, strabismus, deformities of the fingers or toes, polynastia, defective development and irregularity of the teeth and like' (1919: 236).

criminality, vice and unemployment were understood as the expression of inferior primitive psychic and bodily qualities and processes.

The idea of madness as a state of degeneracy was also seen to be expressed *psychic-ally*, through what we might now view as psychological states, as well as *physically* through the signs and symptoms of the decay of the body. For some, according to Emile Kraepelin, degeneracy could even be expressed through one's choice of career or lifestyle. At this time homosexuality was considered a sign of degeneracy manifested through a disease process and it appeared as a psychiatric diagnosis and classification within the textbooks of the time. This particular form of degeneracy, according to Kraepelin, could be expressed through one's choice of employment, 'such as among decorators, waiters, ladies' tailors; also among theatrical people'. He even claimed that women comedians are regularly homosexual (Kraepelin1913: 510). These views were incorporated into governmental strategies, such as eugenics, that argued that the identification, mapping, elimination, segregation and rehabilitation of Otherness (as degeneracy) would allow for the smooth running of the social order.

BIOLOGISM

Sociologists and other humanities scholars who politically want to distance them-selves from what are taken to be biologically reductionist arguments coined the term 'biologism' to refer to arguments that reduce the complexity of human psychological and social life to the biological make-up of individuals and groups. These arguments are also considered *essentialist*, as, again, they reduce the complexity of life to essential components of our biological make-up that are viewed as fixed and pre-given. Diana

---

In this example we can clearly see how madness was seen to wreak havoc on the body causing deformities, irregularities and physical signs that the body was progressing to more 'primitive' forms of behaviour and experience. This view of madness as a form of degeneracy was repeated to explain and reinforce the social positioning of groups who were also considered Other to a white, male, middle-class version of rationality. This included the place of the bodies of children, colonial subjects, the working classes and people with different sexualities. Their bodies were considered potential sites of atavism. 'Atavism' was an evolutionary term used to refer to madness as an expression of biological decay and regression to so-called pre-civilized modes of conduct. Thus, certain groups and individuals were viewed as bearing the seeds of their own destruction within their biological make-up or constitution.

Fuss thus defines essentialism in the following way: 'Essentialism is classically defined as a belief in true essence – that which is most irreducible, unchanging and therefore constitutive of a given person or thing' (1990: 2). We can see in this context why the arguments of Durkheim and his view of cultural inscription were in opposition to many of the views prevalent at the time of his writing. This was not only the case in the psychological and biological sciences, but also in practices of government and regulation. This has led in Turner's view to a suspicion of biological explanations by social theorists and to a lasting commitment to the central role of social processes in the formation of human subjectivities. As he argues, 'any reference to the corporeal nature of human existence raises in the mind of the sociologist the specter of social Darwinism, biological reductionism or sociobiology' (1984: 1).

However, Turner (1984) was writing at the beginnings of what many are now referring to as the emergence of a distinct 'body theory' or area of 'body studies' within sociology. That is, theory that takes the body as a central locus of concern and analysis in relation to broader questions related to power, ideology, technologies, agency and so forth. At this point, what is clear from Turner's position is that what defined the sociological project, as we have seen, was a distance from and even hostility to any engagement with the 'biological'. I have placed the term biological in speech marks as I want to stress that sociologists and social theorists now prefer to use other terms to refer to the body to avoid the spectre of biological reductionism. They also wish to show an awareness that biology is itself a discipline characterized by competing perspectives on the question of how to understand the role of biology in the formation of subjects. Thus there are three interchangeable terms that tend to appear in the literature: the first is the term *corporeality*, used by Turner in the quotation above, which pertains to the body and is a way of referring to the body that does not reduce it to the biological; the second term is *materiality*, which, again, recognizes the material basis of human subjectivity but does not privilege biology as the unique discipline to provide a purchase into this realm; the third term is *somatic*, which again refers to the body but within many perspectives also introduces the concept of feeling or vitality into the body. See, for example, the Introduction, where we encountered the concept of the *somatically felt body*.

THE MATERIALIST BODY

The kinds of perspectives in opposition to which sociology was seen to be defining itself are those that have been characterized as taking the *naturalistic* body as their object of study. Indeed, in his seminal book in the field of the sociology of the body, *The Body and Social Theory*, Chris Shilling devotes an entire chapter to the naturalistic body. He argues that it is this body that has exerted a far greater influence in other traditions than the kind of body that Durkheim, for example, was implicitly formulating within his theories of cultural inscription. He finds the influence of the naturalistic body evidence of the 'power of the biological body' (Schilling 1993: 41), and links this specifically to the rise of sociobiology. This is a contemporary formulation that offers a revision of Darwin's theories, reproducing the idea that there is a biological explanation and basis for human behaviour. As Shilling argues, 'socio-biology begins with an interpretation of current social life – which is often sexist, ethnocentric and factually wrong in other ways – and projects this back onto a mythical history of human societies' (1993: 52). However, although, hopefully, it is now clearer why cultural inscription became one of the dominant traditions within sociology, we need to explore why this move is one that is seen to eclipse or avoid the issue of exactly what we mean when we call for the body to be taken more seriously. A contemporary variant to emerge from this early work on the sociology of the body, as we will see, is one that argues that the implicit body of social theory needs fleshing out. One criticism relevant to this variant is that within cultural inscription models, the material or corporeal body disappears and is replaced by cultural signs and symbols. We will explore this 'socially constructed body' in the next section. For now, I wish to signal that one aspect of bringing the body back into social theory has been a revisiting of the materiality of the body, and a reengagement with the biological sciences as potential allies rather than adversaries. We saw this in the Introduction in Shilling's argument that one uniting principle of contemporary work in body theory is a commitment to exploring the intersection of biological, social and cultural processes in subject formation. As Thrift (2004: 57) cogently puts it, 'distance from biology is no longer seen as a prime marker of social and cultural theory'.

We have seen so far that cultural inscription models moved as far as possible away from biological explanations that were viewed as essentialist, reductionistic and universalistic. In other words, cultural inscription models were based upon an assumption that the idea of a fixed universal human nature that could define subjects for all times, in all places was politically and theoretically suspect. It was seen to position subjects differentially and devalue and dehumanize those who were so positioned according to their bodies rather than the rationality of the white, middle-class male. We can see, therefore, that approaches to the naturalistic body reproduced a dualism that we explored in the Introduction; that is, a separation or distinction between the mind and body that led to certain people being defined by their bodies rather than their minds, which were considered the seat of rationality. We can see, thus, that dualisms not only work on the basis of separation, but are also organized hierarchically. The body within Cartesian dualism is considered a constraining force that ideally should be brought under the control of the mind. The term that is often used within variants of Cartesian dualism is the concept of *will*. Will is a capacity that is linked to a particular form of psychological control that allows the body to be subsumed by the mind (Smith 1992). The body within these formulations is either viewed as machine-like or as the seat of irrationality and emotion. In those who lack the capacity of will the body can potentially override cognition and create mental and physical disturbances. This is what Turner defines as the body produced as 'the location of anti-social desire' (1984: 37). This separation between mind and body, and the hierarchy that is produced on the basis of this split, is often framed through a 'hydraulic metaphor' (Harre 1986). The 'hydraulic metaphor' is one that produces emotion as a physiological state that often gets the better of us. It is the body reacting rather than being acted upon. As Harre illustrates when discussing this dominant version of emotion, the natural and physical sciences reproduce this assumption by viewing the body, and particularly the autonomic nervous system, as the site of the determination of emotion (also see Despret 2004).

THE SOCIALLY CONSTRUCTED BODY

We will now move away from the naturalistic body and consider what other kinds of body have formed the backdrop to approaches across the humanities that reject the essentialism and reductionism of naturalism. The approaches that we will consider in this section have developed the assumption that cultural inscription is the mode through which we become subjects, and have drawn from different perspectives and traditions to make this argument. Within the literature they are often united under the umbrella term of *social constructionism*, and are also often referred to as cultural inscription. As in the latter, the basis of how we become human is linked to the role of social and cultural processes in our formation. Thus, if we want to understand what it means to be human, we need to understand how the body is constructed through symbols, codes, signs, signifying activity and discursive practices. However, although I am using the concept of the body within this context, it is important to understand that this work sought to overcome the dualism between the mind and body. Within the literature, therefore, the concept of the *self* is often used to avoid this assumed split. Thus, Turner (1984) frames this move as one that viewed the self as symbolically constituted rather than biologically determined. We started the chapter by considering one of the key strategies of sociological work on the body that began to make the body more central to social analysis. This strategy was one that sought to recover key sociological figures whose work was seen to engage the body, albeit in implicit or unacknowledged ways. In a similar vein Turner explores how the work of early sociologists such as George Herbert Mead (1934) had already introduced an idea of the self as constituted through symbolic interaction. This is not to say that this work is not without its problems, but rather that it started from the basis that the self is not a unified, unique self, but one that is constructed through our encounters with others. These others are seen to reflect back to us particular ways of imagining and performing self-identity. This self is also often referred to as a 'looking glass self' and illustrates how self-identity is always the expression and manifestation of our incorporation of how we are positioned and responded to by others.

THE MICRO AND THE MACRO

Turner illustrates how variants of contemporary sociological thinking have developed this view of the self as symbolically constituted in different kinds of ways. Turner uses a distinction between the micro and the macro to differentiate work that is part of a social constructionist tradition. The concept of the micro is used to refer to work that focuses upon how this self-construction occurs on a minute level in our encounters and exchanges with others across different social contexts. Thus the sites within more microsociological work that explore this level of symbolic constitution, often focus upon the minute by minute construction of the self within particular conversational settings. This work draws on particular kinds of methods derived from ethnomethodology, such as conversational analysis and discourse analysis (see Potter and Wetherell 1987). Ethnomethodology is a tradition within sociology that explores how social contexts define and set the parameters through which particular kinds of action and interaction are made possible. It is assumed that if you want to understand the self, you need to study language as a human and cultural invention that produces the possibility of particular forms of self-identity. These microperspectives are based upon the insight that humans could only come to know their worlds through social action and negotiation. There is seen to be nothing innate or predetermined about human sense-making activity. Within these perspectives the body becomes a vehicle for the expression of self, but in most cases what is explored are the kinds of talk or accounts that subjects give in particular social contexts. What is reified is conversational activity, and the body is somewhat submerged behind a commitment to the central role of language in constructing human understanding. Microperspectives have also been a central tradition within critical psychology for exploring the 'self' as socially constructed.

These microperspectives might be referred to as *weak* versions of social constructionism. Although the focus is on how cultural codes and symbols construct the body, or what we might term the 'bodyself', the focus is on language as a subjectifying force. The concept of *subjectification* is linked to the work of Michel Foucault, which we will explore later in this chapter, and refers to the processes through which subjects are made, and make themselves into, particular kinds of subject. Within microperspectives or versions of weak constructionism language is the key site through which subjects are made and make themselves. Language creates and forms individual understanding. Culture is made up of a series of texts or narratives that are available as resources through which the individual makes sense of the social world. These cultural narratives are studied or accessed through individual talk that is viewed as symptomatic of these wider texts. Despite the commitment to the body as a socially constructed body, we can clearly see how it becomes hidden or eclipsed by the focus upon language and talk. Although the tradition of social constructionism aims to overcome the problem of dualism in thinking through the body, what we end up with in microperspectives is a view of culture from the neck up. This reproduces the mind–body dualism in a more sophisticated form by inadvertently focusing upon sense-making as a cognitive activity, rather than as a thinking, bodily, felt sense. As Turner (1984) argues what we end up with, brought in through the back door, is a separation of the self from the body, and the body again disappears from analysis.

In contrast to microperspectives the body is seen to play a more central role within macroperspectives united within the tradition of social constructionism. In constrast we might refer to these macroperspectives as *strong* versions of constructionism that are concerned with the relationship between bodies and power. Macroperspectives are aligned with work that has had a prominent place within sociology but that may not necessarily have been explicitly examined for the kind of body it was bringing into social analysis. One of the key proponents of a view of the body as a socially constructed body, integral to macroperspectives, is the French post-structuralist philosopher, Michel Foucault. As Turner (1984) argues, in the work of Foucault, we see a central commitment to a view of the human body as an effect of power and discourse. Now, the collection of work that Foucault produced throughout his life is vast, and shifted and changed from his earlier work on the disciplining of the body through to his later work on self-production contained within the three published volumes of *The History of Sexuality*. There are many useful and accessible introductions to the work of Michel Foucault that are available as secondary reviews of this literature (see the Annotated Guide for Further Reading for outlines). I would recommend that one of these texts is consulted to develop a more engaged understanding of the key concepts that I will outline in the next section to help you understand how he analysed the body as a constructed body.

THE DISCIPLINED BODY

In this section I will concentrate on Foucault's (1977) study of the modern prison system to illustrate the rethinking of power and its relationship to the body that was being developed in this work. Foucault called the type of power that he was illustrating *disciplinary power*. We often think of power as operating in a repressive or prohibitive mode, preventing and constraining action. Power is taken to act upon us, so that freedom, or liberation, is often presented as an over-turning or an over-throwing of power (see Rose 1999). We clearly saw how this view of power was reinforced when, during the invasion of Iraq, 'democracy' was symbolically represented by the toppling of the statue of Saddam Hussain in central Baghdad. Foucault turns this formulation of power on its head, and argues that contrary to our most held or cherished beliefs, power works on and through our actions making possible certain ways of being and doing. He used the concepts of *positivity* and *productivity* to describe the role power plays in producing our becoming. This is in opposition to the concept of power as repression, constraining what we might take to be our 'true' self, for example. Disciplinary power is therefore a form of power that does not prohibit or constrain. Rather it acts on and through an individuals' self-forming practices so that individuals come to want or desire certain ways of being and doing for themselves. It works, Foucault suggests, through the ways in which norms and regulatory ideals become incorporated into subjects' internal forms of self-monitoring and self-regulation. This is achieved not through imposition but rather through their inculcation into particular body techniques and practices. The use of the term *inculcation* as opposed to *imposition* is to stress that if one is inculcated into a set of practices one has to actually *actively* participate. The notion of imposition carries with it the connotation that these practices have been imposed, perhaps beyond one's wishes or will.

Although Foucault was keen to stress how disciplinary power works through the acceptance and active participation of its subjects, he did focus upon a particular institutional context to illustrate his claims. He argued that disciplinary power works most effectively in hierarchical institutions, such as the prison system, where prisoners are living under detailed and often total surveillance. He argued that the organization of power within an institutional practice such as the prison system, works through transforming people's relationships to their own bodies and sense of selfhood. In the context of the prison system, for example, he argued that there are a range of techniques employed that work by transforming the bodies and souls of prisoners. These include strict timetabling, collective training, exercises, total and detailed surveillance (what Foucault described after Jeremy Bentham's original, as the 'Panopticon') and the continual monitoring, assessment and comparison of the prisoners in relation to particular norms of behavior and conduct. The following quotation, taken from an extract from Foucault's work *Discipline and Punish*, brings together some of the themes I have been discussing:

> Procedures were being elaborated for distributing individuals; fixing them in space and time; classifying them; extracting from them the maximum in time and forces; training their bodies; coding their continuous behavior; maintaining them in perfect visibility; forming around them an apparatus of observation, registration and recording; constituting on them a body of knowledge that is accumulated and centralized. The general form of an apparatus intended to render individuals docile and useful. By means of precise work upon their bodies, indicated the prison institution, before the law ever defined it as the penalty par excellence. (Rabinow 1991: 214)

The concept of the *docile* or *disciplined body* has been taken up in many studies of the body across social theory and feminist work. It is seen to be useful as it presents the body as malleable, as an unfinished entity that can be sculpted, moulded, altered and transformed. It also draws attention to the ways in which social norms can become internalized and operate through our own self-forming and self-regulating practices. It also draws our attention to how these practices can become engrained and embodied in such a way that they appear automatic and natural. Foucault rejected the idea of a universal self, arguing that our practices, habits, desires and beliefs are produced, and are not simply the expression of a pre-existing self. Power is therefore *constitutive* rather than repressive. This work is very close to the model of cultural inscription that, as we have seen, has historically been centrally important

Case Study

Susan Bordo (1997) and Sandra Lee Bartky (1997) have provided illuminating Foucaudian analyses of eating disorders, and particularly of how we might understand the high incidences of anorexia nervosa amongst young women. Eating disorders within the psychological and biomedical sciences are viewed as pathologies. The term 'pathology' is used to denote experiences that are considered abnormal, and usually signify as symptoms of disease and illness. In the case of eating disorders they are considered symptoms of illness, although there remains speculation over whether the illnesses are psychological, physical or some kind of amalgamation of the two. Both Bordo and Bartky reject the concept of pathology for explaining the incidence of eating disorders amongst women, and instead link femininity and eating disorders through a notion of the *disciplined* or *docile* body. They focus upon practices that they both argue have been aligned with femininity to how the discipline of sociology defined its project. The ramifications of some of the concepts that Foucault developed have been far-reaching. They have been appropriated by critical psychologists (Blackman and Walkerdine 2001; Rose 1989, 1996, 1999; Hook 2007), postcolonial writers (Ahmed 1998, 2004; Bhabha 1994), feminist scholars (Bordo 1993; Bartky 1997; McNay 1992) and others across the humanities who have rejected the essentialism of the naturalistic body and turned instead to a socially constructed body. Foucault's work has been central to the formulations of Bryan Turner (1984), who is one of the leading proponents of body theory within sociology and social theory. However, this work is not without its critique, and although providing a platform for the reinvigoration of work on the body within social theory, has also brought with it a number of problems and paradoxes. These problems and paradoxes form a thread through some of the work that we will explore in later chapters.

Shilling argues that the body within social constructionist traditions is a body that is 'shaped, constrained and even invented by society' (1993: 70). The body is stimulated into being, rather than repressed by brute force so that its physicality or materiality becomes the raw material for cultural processes to take hold. The body is there, but as many scholars have argued becomes 'inert mass' (Shilling 1993: 80), or a passive effect of cultural discourses. Others have argued that although Foucault coined the concept of the 'docile body' to describe the relationship of bodies to power, it is actually the mind rather than the body that is appealed to. Again, we end up with a dualism between the body and mind, albeit in approaches that privilege social processes as determining the thinking body. As Woodward argues,

through cultural representation. These include 'self-improvement', dieting, beautification practices, such as cosmetics and cosmetic surgery, practices of adornment such as fashion, and the myriad of practices available to women to transform the female body into 'a body of the right size and shape' (Bartky 1997: 136). They argue that the *anorexic body* is simply an exaggeration of these so-called normal feminine practices in the pursuit of a perfect, slender body. It is not that we can distinguish practices of femininity and those of the anorexic through a differentiation between the normal and the pathological. Rather, they both argue that there is a continuum between 'female self-improvement' and what Bordo claims we see in the anorexic with 'their inscription in extreme or hyper literal form' (1997: 97). Female 'self-improvement' is considered a mode of disciplinary power.

once the body is contained within modern disciplinary powers, it is the mind, which takes over as the location for discursive power. Consequently, the body tends to become inert mass controlled by discourses centered on the mind (which is treated as if abstracted from an active human body). This ignores the idea of disciplinary power as 'lived practices' which do not simply mark themselves on people's thoughts, but permeate, shape and seek to control their sensuous and sensory experiences. (Woodward 1997: 79)

AGENCY AND THE BODY

This touches on one of the main critiques of the socially constructed body that have become a platform for more recent work on the body across social and cultural theory. It also takes us back to some of the problems with the early sociological work that brought to the fore the role of social structures in the formation of human subjectivities. This critique is usually framed as the 'problem of agency'. As Shilling (1993: 81) argues, 'the body is affected by discourse, but we get little sense of the body reacting back and affecting discourse'. Within this work, there is little sense, then, of how bodies might protest, speak back or simply refuse to participate in the workings of disciplinary power. There is assumed to be a tight fit between what is often termed the *social body* – that is, the body that is constructed within ideological and social processes – and the *physical body*. The social body is seen to mesh tightly with the physical body to the extent that they are seen to be copies or mirrors of each other. The term that is often used in the literature is 'homology', meaning that there is no separation between self and social identity. Because of the homology assumed within the literature a problem is set up: how to theorize agency? The concept of agency refers to the individual's capacity to resist, negotiate or refuse the workings of disciplinary power, for example. This is usually spoken of as the structure/agency dichotomy and is often the starting point for contemporary work on the body across the humanities that are attentive to the body's capacities for agency. We will explore different responses to the structure/agency dichotomy throughout the chapters to come.

Although approaches to the socially constructed body recognize the body's capacity for malleability they are now regarded as socially deterministic. For many scholars the move to this body as an alternative to the naturalistic body is a move from one form of determinism, *essentialism*, to another: *discourse* or *social determinism*, which is seen by many to be a more sophisticated form of essentialism. Let us consider the following quotation from Diana Fuss that brings together some of the key problems that centre on the problem of agency presented by the socially constructed body:

[It is] articulated in opposition to essentialism and concerned with its philosophical refutation and insists that essence is itself a historical construction. Constructionists take the refusal of essence as the inaugural moment of their own projects and proceed to demonstrate the way previously assumed self-evident kinds are in fact the effect of complicated discursive practices... What is at stake for a constructionist are systems of representations, social and material practices, laws of discourses and ideological effects. (1990: 2)

We can see from this that social constructionism is articulated in opposition to naturalism or essentialism. We have explored the historical antecedents of this opposition in relation to the project of sociology throughout this chapter. One form of determinism is replaced by another, with the question of the materiality or corporeality of the body sidelined by constructionism in its rejection of biological reductionism. What we have is a dissolving of the body into what Denise Riley terms 'the clouds of the social' (1983: 41). The body and its biological potentialities are foreclosed. The body is presumed to be passively written upon, so that the 'dynamic nature of the body' (Shilling 1993: 104) is silenced and ignored. Naturalism and social constructionism, as Shilling makes clear, are therefore opposing forces linked by reductionism. We saw in the Introduction how contemporary work within the sociology of the body is grappling with how it is possible to reintegrate or realign these two perspectives, in such a way that we can analyse the body as a 'phenomenon that is simultaneously biological and social' (Shilling 1993: 100). Towards the end of this chapter we will consider a contemporary study within sociology on sleep that has attempted to do just this (Williams 2005). What I want to do in the next section, however, is focus upon another critique of the tradition of the socially constructed body that relates to the assumption of the body as inert mass. This relates to what many refer to as the 'somatically felt body', and introduces an aliveness or vitality into a body that can be reduced neither to physiological processes, nor to the effect of social structures.

THE SOMATICALLY FELT BODY

Within Foucault's study of the prison system we saw how repetition and repeatability are central to the workings of disciplinary power. Prisoners, for example, are exposed to practices that work through repetition, with the intention that individuals take on the responsibility for monitoring and regulating themselves. Similarly, in the case study on eating disorders, women are seen to be continually invited to judge and evaluate themselves in relation to normative feminine ideals. Bartky (1997) and Bordo (1997) argued that women are exposed to cultural ideals and regulatory images of the female body that are repeated across a diverse range of practices. In this section we will give consideration to two issues: a study of practices that work through such repetition and an aspect of the body that Foucault disregarded or did not give thought to: what I term 'the somatically felt body'. The somatically felt body has aliveness or vitality that is literally felt or sensed but cannot necessarily be articulated, reduced to physiological processes or to the effect of social structures. The study I am going to refer to is one carried out by a historian who is examining some of the paradoxes raised by his own experience of taking part in military drills in the army. This study is interesting as it speaks to some of the key criticisms that have been made of Foucault's concept of the docile body; that is, as we have seen, that the body becomes inert mass and it is the mind that becomes the target and object of disciplinary power. This study of military drill makes visible aspects of corporeality that are missed by Foucault's study, and, as we will see throughout the book, are the aspects that are becoming a central focus of contemporary sociology and social theory.

I want to start by quoting in length from the study, which introduces a concept of *muscular bonding* to refer to the kinds of affective or emotional experience that are often produced when people move together rhythmically in time. This might be in forms of dance or in the example that the author gives of moving the muscles rhythmically in army drill.

> Marching aimlessly about on the drill field, swaggering in conformity with prescribed military postures, conscious only of keeping in step so as to make the next move correctly and in time somehow felt good. Words are inadequate to describe the emotion aroused by the prolonged movement in unison that drilling involved. A sense of pervasive well-being is what I recall; more specifically, a strange sense of personal enlargement; a sort of swelling out, becoming bigger than life, thanks to participation in collective ritual. (McNeill 1995: 2)

This example suggests that one of the aspects of corporeality that bind people together is a sense of cohesion which is experienced through the body as an expansive feeling. The concept of muscular bonding that McNeill develops refers precisely to what he terms this 'emotional affect of rhythmic movements and gestures' (1995: 5). This felt, visceral sense of feeling in tune with others is one that has a long tradition within work in anthropology on ecstatic cultures (see Lewis 1971), which tends to focus upon trance states that are brought about through repetitive, ritualistic practices marked by 'prolonged or heightened exertion' (McNeill 1995: 8). For some, these experiences are viewed as pathological and are 'Other' to normal psychological functioning. One such evaluation of these states was made by a famous British psychiatrist, William Sargeant (1967), who was motivated by a fascination with religion and phenomena that, in the cultures he studied, were experienced as signs of a divine, sacred world. These included the healing methods of 'witch doctors' in Ethopia, Kenya, Zambia, Nigeria and Dahomey; fire walking in Fiji; temple drumming and dancing in India; transitional religious practices in Brazil; Voodoo in Haiti, and Revival meetings across North America. People experienced these states as spontaneous experiences of being possessed or taken over by spirits. This experience of possession was felt in and through their bodies in a range of sensory modalities including motor dissociation, contortions and tics, trembling, tingling in the hands and other body parts, catatonia, fainting, trances, stupor, collapse and feelings of heat, lightness, heaviness and so on and so forth. Sargeant drew parallels with political techniques of conversion in his preoccupations with Communist, Fascist and Nazi revolutions across Europe, and argued that conversion, both religious and political, could be explained by a physiological mechanism, an abreactive reaction of the brain, brought about by rhythmic and repetitive behaviour (1967: 171).

This is a reductive explanation typical of the kind of biologically essentialist approach that we explored in relation to the naturalistic body. It is assumed that these experiences can be explained solely by a physiological mechanism in the brain (abreaction). He dismisses the practices that he examines by aligning them to primitivism and therefore argues that they have little to tell us about practices that exist in Western cultures that he does not consider exceptional or abnormal. Military drill or dance are two such examples that McNeill, however, suggests induce a feeling of 'rhythmic kinaesthetic stimulation' that is part of the glue or cement that binds individuals together within the group (McNeill 1985: 7). McNeill's history of muscular bonding shows how the 'muscular, rhythmic dimension of human sociality' has a long history (1985: 156). He charts its importance in animal societies, in the community binding festivals of dance in small communities, in religious ceremonies and in politics and war. In relation to politics and war, it is Hitler's use of muscular bonding during the Third Reich that is seen to exemplify the affective basis of practices that bind people to each other and induce a sensation of solidarity or 'fellow-feeling'. Hitler mobilized the use of marching and other forms of repetitive drill on a grand scale in the huge rallies, such as the Nuremburg rally, that were filmed by the German artist Leni Riefenstahl (1934) and used as propaganda. These films are now available for viewing and provide a disturbing account of the role of muscular bonding in politics and war. We will explore some of these practices in Chapter 2 where we will consider 'communicating bodies' and the role of affect and emotion in bodily practices.

McNeill suggests that 'repugnance against Hitlerism has discredited rhythmic muscular experiences of political and other sorts of ideological attachment' (1995: 151). He suggests, then, that we have not been keen to explore this gestural, muscular level of communication, preferring to see it as an abnormal or pathological phenomenon that occurs in what are deemed to be more primitive societies, or in those who are seen to have lost their will and submitted to the will of a charismatic leader – in those, in other words, who have lost the capacity for rationality and subsequently become defined by their bodies. This mind–body dualism, as we will see throughout the book, is entrenched and makes an appearance in many guises. However, although not wanting to reduce the affective glue that might bind people together to muscular bonding, it is a concept that introduces an aliveness or viscerality into the body. It is not just inert mass, but reacts back, responds, often at a level that is felt through the body but might not easily be open to articulation. One of the problems of cultural inscription or the socially constructed body is precisely the way in which the body is viewed as passively written upon and does not seem to have any energy or creative motion. As many people are now arguing, the body that needs to be brought into social and cultural theory must be one that is also enhanced, modified and managed through a recognition of the importance of a register of feeling, affect and emotion (Tamborinino 2002; Thrift 2004). We will discuss these arguments in more detail in Chapter 2. As we can see, then, ironically the move to social determinism further displaces a sense of exactly what kind of body we want to make central to sociology and social theory. The problem as Turner (1984: 248) suggests is to overthrow a 'number of perennial contrasts' between, for example, structure and agency, mind and body, nature and will and the individual and society and to offer solutions that are neither deterministic nor view the body as somehow existing prior to social and cultural processes. These are the tensions and paradoxes with which the range of studies, traditions and perspectives that we will review in later chapters of the book are trying to grapple.

THE SLEEPING BODY

We will finish this chapter with a review of a contemporary sociological study of the 'sleeping body' that is committed to exploring the intersection of the biological, social, cultural, psychological, physical and economic processes in the production of sleep. The aim of this study is to guard against reductionism of any kind, and, in the words of the author, sleep 'is a complex, multifaceted, multidimensional phenomenon that cannot be reduced to any one domain or discourse, be it biological, psychological, social or cultural' (Williams 2005: 169). As Megan Brown (2004) argues, sleep has literally become big business in modern corporate culture, with many work organizations turning increasingly to a burgeoning array of sleep consultants to improve the well-being and productivity of their workers. This might include the provision of 'sleep stations' for employees to take sanctioned 'power naps' or the provision of corporate training to help employees to enhance their own sleeping patterns beyond the office walls. Brown reviews the huge range of self-help books, newspaper reports, magazine articles, workshops and consultants who offer advice and practices in this micromanagement of sleep. In the United States 500 companies, among them American Express, Ford and AT&T Network Systems, have commissioned such programmes to date.

As Brown and Williams both discuss, sleep has increasingly become medicalized with a whole branch of medical science devoted to sleep pathologies and disorders. This branch of science, which intersects with the psychological sciences, also produces a vast array of knowledge and practices on sleep hygiene that has culminated in what Williams (2005) refers to as a sleep industry, which is itself supported by the pharmaceutical industry and is based upon the measurement, classification and diagnosis of sleep, as well as the provision of a range of practices and prescription hypnotics to address what are being identified as a range of new sleep pathologies and disorders. The discourses produced by this industry include the identification of both the benefits of sleep and the dangers of sleep deprivation. Although Brown suggests these discourses are not new, what is new, she proposes, is 'the connection between corporate policy, management strategy, and sleep-related medical and self-help advice' (2004: 174). She identifies the parallel between the medicalization of the human body that is integral to sleep medicine and the potential for its micromanagement by employers and employees. Her argument, following Foucault, is that 'workers can be governed – and learn to govern themselves – even through the basic, mundane bodily phenomenon of sleep' (Brown 2004: 175). What does it mean, then, that a so-called basic bodily need for sleep can be micromanaged in order to optimize performance and productivity in the workplace?

Williams and Brown both allude to the way in which sleep is often viewed as one of the most private, intimate and personal activities that we carry out. Sleep is often assumed to be asocial or non-social and therefore to be of little or no concern to sociologists and social theorists. It has been neglected by sociology, and, as Williams testifies, there is little in the discipline to address this complex practice. As we have seen, the practice of sleep, or the sleeping body, is dominated by medical science, although the institutional and social patterning of sleep is becoming a concern for corporate management. Williams relates his aim to make the sleeping body a central concern for sociology part of a broader question of what kind of body we wish to bring back into social theory and how. He suggests that the sleeping body is important as it lies between a number of dualisms that social theory is attempting to think against. These include the voluntary and the involuntary, the purposive and the non-purposive, the personal and the impersonal, the biological and the social and the universal and the specific (Williams 2005: 4). It is a 'between state' that has been aligned with states such as intoxication, hypnosis, anaesthesia and stupor. It is a complex practice that cannot simply be equated with closing our eyes. What kind of body, therefore, is Williams suggesting is important for social theory in his consideration of sleep as a complex practice?

The sleeping body is not simply a biological body governed by a need for sleep set by the body's own internal clock or circadian rhythms. This is the body that is the norm within medical science, which devotes itself to those who find it difficult to sleep, or whose sleep is interrupted due to disorders such as sleep apnea. He shows how sleep difficulties have a long history that can be charted throughout literature, for example, and reveal the complex practices that have been adopted to resolve such difficulties. These include praying, dream interpretation, music, meditation and acupuncture. He also illustrates how the sleeping body cannot be determined by biological rhythms by exploring the radically different ways that sleep has been organized and institutionalized throughout history. Sleep exhibits a wide variability when we compare the varying ways in which it has been arranged, problematized and organized. It 'is no more a biological given, but a historically variable phenomenon; an important indicator or index, in fact, of social order and change' (Willliams 2005: 65). The historical organization and patterning of sleep is an important barometer of the shift and change towards the idea of the separated body that was mirrored through the emerging practice of segmented sleep. This introduced strict boundaries between the self and other that were institutionalized through new practices of sleeping in private quarters away from animals and non-intimate others. It came increasingly to signify social status, power relations and privilege, for example.

EMBODIMENT

The kind of approach to the 'sleeping body' that Williams is advocating is what he terms an *embodied* perspective. This is a perspective that would go beyond viewing sleep solely as a biological or physical process yet at the same time does not dismiss the contribution of the materiality of the body to its social patterning and modification. Sleep is 'embedded' within a network of social roles and relations (Williams 2005: 97) and exhibits a wide variability and flexibility. The contribution of the kind of body that Williams seeks to bring back into sociology and social theory is a body that shows how the distinction between nature and culture is in practice impossible to untangle. Nature and culture are not two separate distinct entities, but rather exist in a complex *relationality* that is contingent and mutable. What we might identify as nature – the simple fact of sleep – actually, through the close examination of Williams' embodied perspective, becomes a fact of convention. Within this perspective, which starts from a rejection of separation and dualism, the sleeping body is a body that is never defined solely by physical needs, nor is it separate from the complex processes that we might define as social, cultural, economic and so forth. As Turner (1984) suggests, this is a view of biology as a socially mediated phenomenon. The work of Williams goes someway towards both introducing subject matter into social theory and sociology that has historically become the province of the natural and biological sciences. It also makes important moves towards rejecting the idea that nature and culture, for example, exist as separate entities that somehow interact. Rather, they produce each other in such a way that it is impossible to disentangle one from another. We will explore in later chapters many different perspectives that also start from this position and introduce different concepts for thinking through the complex *relationality* that the body presents for social theory.

Conclusion

Within this introduction to body concepts within sociology you will now be more familiar with some of the issues and debates in relation to which the body is defined. As you will also be aware, when we talk about the body, or call for its reappraisal, it is never a singular body. The very notion of a separate, singular body is itself a historical construction that is part and parcel of the problems of thinking through the body for social theory. We will examine the emergence of this particular notion of a separate, singular body that is 'affectively self-contained' (Brennan 2004) in Chapter 2. When, in the 1980s, sociology and social theory began to make the body more explicit and to bring to the foreground what had been lying dormant in the background, it also brought with it a number of key concepts that we will be exploring throughout the book. These include some that we have touched on in this chapter: *embodiment, corporeality, affect, emotion, materiality, discipline, process, practice and technique*. These should now be more familiar to you, and will appear in different ways throughout the chapters ahead. Thus, as many sociology scholars have argued, the body is central to many of the paradoxes that govern sociological thought; the question still remains what kind of body or bodies will enable us to think through these paradoxes in new and exciting ways. We will begin this reflection in Chapter 2 by exploring how the concept of *embodiment* is a distinctly different paradigm to the concept of *social influence*. We will explore this in relation to what I will term the 'communicating body' and particularly in relation to communication that cannot be captured by the concept of cognition.

2 COMMUNICATING BODIES

INTRODUCTION

In this chapter we will examine a number of key body concepts that will enable you to differentiate the paradigm of *embodiment* from the concept of *social influence*. We ended Chapter 1 with a study by a contemporary sociologist considering the 'sleeping body' from an embodied perspective. As we saw, this perspective refuses the idea that the biological and social, the natural and the cultural, exist as separate entities. Embodied perspectives start from the position that nature and culture are not separate, pre-existing entities. If we have this as our point of origin we do *not* have to explain how these two entities come together or influence each other. In contrast, the concept of *social influence* assumes precisely the existence of separate realms that somehow *interact*. The place where the two realms are said to come together is what is known as the 'interaction effect'. Denise Riley (1983: 28) suggests that the use of the idea of an 'interaction effect' does not resolve the problem of how to think the intersection of the biological and the social. In fact, it sets up some of the very paradoxes and tensions that sociologists of the body have been trying to overcome and avoid. Thus, as Shilling (1993: 12) argues, the conception of the *natural body* that underpins the concept of social influence produces the biological as a fixed realm of determinate processes, rather than as an 'unfinished' phenomenon. The natural is seen as a biological base upon which social influences processes can only take hold in very specific ways. We encountered this approach to the biological in the Chapter 1 in the section that explored the emergence of the *naturalistic body*.

SOCIAL INFLUENCE

The concept of *social influence* brings into being both a concept of the *naturalistic* body and a conception of the social that refers to those processes most likely to influence the biological in a fairly peripheral fashion. As we have seen, it is not just that the natural and the cultural or the biological and the social are separate but that, usually, the natural is taken to refer to a realm that is more fixed, and the cultural to a realm that is subject to change. Thus, the concept of social influence assumes a particular conceptualization of the body and its social environment. That is, that the natural body refers to a more fixed realm made up of a static, invariant set of characteristics that predispose persons to particular forms of thought, behaviour and conduct. The social is seen to refer to a realm of cultural processes that are more fluid and contingent. However, they are seen to 'interact' or come together in a particular way. That is, that the body of the individual (made up of particular characteristics) sets limits on their interaction with the 'social'. Thus, the natural body constrains or delimits how the 'social' can impact or impinge upon the individual. The body, therefore, can be affected by social relations, but when such an 'interaction' occurs it is usually seen to be a relationship of *distortion* (Shilling 1993). Shilling therefore argues that this concept is often used to describe women's relationship to the social. This is explored in the case study presented here in a little more detail, so that it is clear how the concept of social influence is taken to operate.

BECOMING (HORSE–HUMAN)

One of the key focuses of this chapter will be on the extraordinary, the apparently inexplicable, the anomalous, and work that tends to be kept in the background as it threatens some of the implicit and often explicit formulations of the body that have entered social theory from the paradigm of social influence. We will start by considering an experiment within early psychology that confounded the experimenters and led to the belief that 'Clever Hans', a Russian horse, might have

Case Study

In a review of some of the feminist approaches to body image and eating disorders that have been written by prominent feminists such as Susie Orbach and Kim Chernin, Shilling argues that, although attentive to the way in which cultural representations mediate women's relationships to their bodies, there is a problem in what is assumed to be *natural* about women's bodies. Thus, the female body is taken to be affected by social relationships, but these relationships are taken to disrupt women's *natural* shape and size, for example.

As he argues, it is assumed that 'women can become alienated from their physicality' (1993: 67). This approach to the diversity of female bodies forms the basis of reality television shows such as *How to Look Good Naked* and has also been mobilized in the Dove 'Real Women' advertising campaigns, where women are viewed as having a distorted relationship to their physicality and body image. This *distorted* relationship is re-educated through engaging in various techniques of self-production and transformation (with the help psychic abilities. Hans the horse was owned by a Russian aristocrat Wilhelm Von Osten who firmly believed that animals possessed an equal capacity for intelligence with humans. On this basis he attempted to teach a cat, horse and a bear to do simple arithmetic. It seems that with Hans the horse Von Osten had found a test case for his theory. Hans appeared to be able to solve fairly complex multiplication puzzles by stamping his hooves. In 1904 a commission of thirteen people led by Carl Stumpf, who was the director of the Berlin Psychological Institution, was assembled to judge and evaluate Von Osten's claims. The members concluded that either Hans was possessed of an exceptional intelligence and/or had psychic abilities. It is no surprise that this experiment was revisited as it presents some startling claims for horse–human relationships. Although, as we saw in the Introduction, work on the body within sociology has begun to explore horse–human relationships in novel and exciting ways (Game 2001), the idea that a horse could solve arithmetic and even tell the time threatened prevailing psychological theories about intelligence and cognition.

Hans became a test case in experimental psychology for the problem of social influence. As we have seen, within sociology and social theory the problem of social influence is seen to be one that sets up a separation and distinction between nature and culture and the biological and the social. Within experimental psychology social influence is often conceived as a kind of bias or error. One example of this, to which the study of Hans the horse has been linked, is the idea of an experimental effect. This concept is used to refer to experimental artefacts that are seen to be artificially produced by the experiment. Many psychological experiments are therefore said to

---

of cultural intermediaries of course). Although these strategies problematize the 'slender body' ideal and encourage women to accept their body shapes, the key problem is what is being identified as *natural*, and how the social is taken to be a distorting or alienating factor. This is somewhat different from the approaches to body image and eating disorders that argue that a woman's relationship to her body is *not* simply one of distortion, but rather is one that is constituted through the workings of disciplinary power for example (Bartky 1997; Bordo 1997). We explored such an approach in Chapter 1. This does not make a claim that we can clearly identify and know what might be *natural* if only women were able to separate themselves from the workings of power and ideology. This might seem a moot point but it is crucial to work in a more embodied perspective that does not make grand claims about what is natural and what is cultural. This will become clearer as the chapter progresses.

be confounded by an 'experimental effect' as subjects of the experiment are seen to produce the answers that they think the experimenter wants. This introduces a self-serving bias or error into the experiment. The problem of social influence conceived in this way has led experimental psychology to frame its study through a concern with how to eliminate or eradicate so-called experimental bias (Rosenthal 1966). Bias or the compliance of experimental subjects with the wishes or demands of the experimenter are viewed as 'parasitic supplements that seriously contaminate the purity of the experiment' (Despret 2004b: 118).

Although this is one way that the study of Hans has been discussed and framed within experimental psychology, it also raises some interesting questions for how and what kind of body we might seek to introduce within sociology and social theory. I will discuss this in relation to the problem of communication, or what I am going to term 'the communicating body'. The work of the French philosopher Vinciane Despret (2004b) is useful as she considers the case of Hans the horse in the context of what it might mean for social theory to consider and analyse the concept of *becoming*. We encountered this concept in the Introduction within another examination of horse–human relationships. The concept was used to refer to the ways in which the relationship between KP the horse and its owner was not one of separation but rather a mutual *relationality* that produced the possibility of their attunement with each other (Game 2001). The concept of *becoming*, like the paradigm of embodiment, refuses the idea of separation; in this case, between the self and other: human and non-human. In order to illustrate how the concept of becoming troubles the foundation of separation upon which 'social influence' is based, Despret turns to the work of the experimental psychologist Pfungst (2000). Pfungst had reconsidered the conclusions of the Berlin Psychological Institute in 1907 and had come to some rather different conclusions about the relationship between Mr Von Osten and his horse Hans. These were published in his book, first translated into English in 1913, *Clever Hans: The Horse of Mr Von Osten*.

In this book Pfungst comes to some rather different conclusions about the relationship between Mr Von Osten and his horse Hans. He argued that Hans was indeed clever, but that this cleverness was not linked to an exceptional ability to solve arithmetic. Rather, the horse was able to read subtle, minimal bodily clues given by Mr Von Osten that he was not aware he was communicating. In other words, it was not that Mr Von Osten was deliberately attempting to deceive the experimental community as he was not aware of the communication that was happening between himself and Hans. Pfungst concluded that 'unintentional minimal movements (so minimal they had not been perceived until now) are performed by each of the humans for whom Hans had successfully answered the questions' (Despret 2004b: 113). This attunement was not linked to the possible special relationship between

Mr Von Osten and his horse as Hans was also able to read minimal cues from other experimenters who might very subtly change posture, or produce a particular gesture when Hans's tapping with his hooves corresponded with the correct answer. Despret concludes from this re-examination by Pfungst that 'who influences and who is influenced, in this story, are questions that can no longer receive a clear answer' (2004b: 115). She argues that this shows the limits of models of social influence that rely upon a clear and distinct separation between entities: self and other, horse and human. She argues that the case of Hans makes visible the capacity of horse and human to transform each other to such a degree that they are affected and affect each other: 'not only could he read bodies, but he could make human bodies be moved and affected, and move and affect other beings and perform things without their owners' knowledge' (Despret 2004b: 113).

The capacity of horse and human to *become* together brings into existence different conceptualizations of the body that do not rely either on singularity or separation. Despret equates this to a 'being-with' that is similar to Game's (2001) conception of connectedness or attunement that we explored in the Introduction. These concepts need a different paradigm to social influence in order to explore this interconnection or intersection. Before we explore this in more detail I want to consider more closely how the body that enters Despret's formulation of *becoming* is radically different to the usual way in which the body enters discussions of minimal communication. This area of research is usually known as the study of non-verbal communication or body language. Within this tradition, the study of the body is central and not displaced by a focus upon cognition. Indeed, non-verbal communication refers to the realm of communication that happens beyond language and conscious deliberation or reflection. However, as we will see, within traditional studies of body language that originate within the psychological sciences the body enters in a very particular way. If we turn, therefore, to the psychological sciences, exactly what kind of body is taken to communicate non-verbally?

BODY LANGUAGE

We all in one way or another, send our little messages out to the world. We say 'Help me, I'm lonely. Take me, I'm available. Leave me alone, I'm depressed.' And rarely do we send our messages consciously. We act out our state of being with non-verbal body language. We lift one eyebrow for disbelief. We rub our noses for puzzlement. We clasp our arms to isolate ourselves or to protect ourselves ... The gestures are numerous, and while some are deliberate and others are almost deliberate, there are some that are mostly unconscious. (Fast 1971: 17)

Non-verbal communication or body language is usually framed within the psychological sciences as involving a mismatch between words and feeling. The concept of *body leakage* that circulates within this literature is a term that refers to a person's feelings that might be at odds with what they are saying or doing. Fast (1971: 1) recalls a classic example where a 'young woman ... told her psychiatrist that she loved her boyfriend very much while nodding her head from side to side in subconscious denial'. Thus, it is assumed that non-verbal communication is produced from a realm of bodily experience that is more authentic and honest. Experts on body language thus claim to be able to identify moments where bodily signs and codes reveal the *truth* of how the person is feeling in spite of their apparent words or actions. This approach to the body within the psychological sciences assumes a contrast between the authentic and the manipulated and the honest and the deceptive, where body leakage is judged according to the extent to which the person is revealing their feelings which might be at odds with what they are saying or doing.

This makes a number of presumptions about self-performance that we can also find within sociological studies of the body. The first is that self-performance is subject to forms of emotion management, where it becomes a key site for the regulation of feeling (Hochschild 1983; Goffman 1959). This is a *dramaturgical* model of the self that explores our success and ability to manage the impressions we give to others. This is a 'performing self' that creates and manages its own image. Turner (1984: 112) links the emergence of this 'self' to a growing 'ethic of managerial athleticism' that is required by political actors to be successful. Arlie Hochschild (1983, 1994) has also explored how certain industries, such as the service and airline sectors, support, extend and encourage employees to perform, manage and regulate impressions and feelings in their relationships with consumers. For example, this might be by being

Case Study

One of the central concepts popularized and validated across reality television is the concept of 'faking it'. The concept of 'faking it' refers to those moments when the participant on the show is shown to be faking it, to be performing or acting for the camera rather than expressing how they are really feeling. This is usually framed as 'just being real', 'being me', or 'being true

to myself'. These ideas have been popularized in the staging of reality television shows such as *Big Brother*, which provide a micropsychological laboratory through which participants can be judged and scrutinized for signs that they are 'faking it'. These signs are often revealed by expert psychologists who expose moments of deception through the unwitting gestures and bodily friendly and smiling. Thus, although the concept of body leakage assumes that there is a realm of authentic feeling, it is difficult to determine what is authentic and what is managed when we consider the presentational strategies that are required to be a successful employee or social actor. This is what Goffman referred to as 'impression management' and assumes an apparent 'knowingness' about the strategies we might use to accomplish certain ends.

Turner (1984) identifies Goffman as one of the sociologists alongside Foucault and Mead (both of whom we explored in Chapter 1) who made the study of the body integral to their explanations of how the workings of power, ideology and social processes were effective. Shilling (1993) argues that Goffman makes central to sociological analysis how the body and its management is integral to the successful management of social encounters. However, although Goffman explored how the body is a key vehicle for self-expression, he also presumed a split between the self and the body. Thus 'knowingness', or the capacity to manage and manipulate the expression of feeling, is linked to a set of strategies of self-presentation purposively carried out by a *public self*. This is a self that is aware of the impressions it makes in relation to others, and is skilled at manipulating gesture in order to present itself in a particular light. Goffman is part of a microsociological tradition that assumes that the 'self' differs in the kinds of competences it has developed. It is these competences that make the manipulation of space, distance or a sense of status or authority possible and probable. These competences are usually framed through the concepts of *self-control, will* or *insight*. This aligns the competences of the presentational self with the possibility of deception, and has produced a volume of literature on 'body tells' (Collett 2003), exploring how politicians and other public figures deliberately use body language to emphasize, persuade and even deceive their audiences.

---

expressions 'caught' by the camera at moments when it is assumed that the subject's capacity to perform has begun to develop cracks. These are the moments of 'unmasking' that viewers, other participants and expert psychologists are so keen to observe. This is the subject 'laid bare' and read or decoded through what is often presented as a universal non-verbal language, enacted through pauses, hesitations and intensities. These are revealed through the articulations of musculature and bodily expression usually aligned to the action of the autonomic nervous system. In other words, this is the moment when what is taken to be the *natural body* radiates through the *social body*.

However, much of Goffman's work focused upon the limits of the 'presentational self': of what happens when presentational work fails or breaks down resulting in 'losing face' or the experience of embarrassment, shame or humiliation (Probyn 2005). As Turner (1984: 111) argues, 'performances are threatened by the possibility of perpetual failure'. This ties the microsociological work quite closely to work in the psychological sciences which looks at how although the bodily expression of feeling can be brought under conscious deliberation there is another realm of expression that is often non-conscious and involuntary. This realm and its disclosure is usually understood as *masking*, where the so-called truth of feeling is almost always revealed through very subtle signs and cues of which the subject is not aware. This is constituted as a realm of *authentic* feeling that is often presumed to be the expression of a self that pre-exists and is separate from social influence processes. This is a model of 'affective self-containment' (Brennan 2004) that assumes that there is a 'self' that is separate from others and that is natural and pre-social. This is rather different to what also might be revealed by this register of minimal bodily communications if we refuse the concept of the separate, singular body. We explored such an interpretation in Despret's account of the case of Hans the horse and what a refusal of separation might disclose or reveal about *becoming*. That is that the mixing and interconnection between self and other does not reveal an authentic separate realm but rather the capacity we all have for being affected and affecting the other. Thus, this unmasking of the 'bare human being' (Fast 1971: 65) might also reveal quite the opposite if we work from a more embodied perspective. That is, the extent to which the boundaries between self and other are permeable, and that even our attempts to manipulate the surface of our bodies do not provide 'an immovable barrier between ourselves and the outside world' (Shilling 1993: 23).

However, the concept of *becoming* and the more embodied perspective towards which it points present a language of feeling that is at odds with the idea of the 'authentic self' that is so embedded in our everyday language of the self and the body. The concept of the 'authentic self' relies upon the concept of social influence, where social processes are seen to mask, hide, manipulate or cover the realm of true, authentic feeling. When these processes are revealed or lifted the true self is seen to emerge. It is, then, rather difficult to imagine the different kind of language of feeling that might emerge with the concept of *becoming*, but as we will see later in this chapter, a language of connectedness rather than separation has already existed, and indeed still exists at the margins of our emotional vocabularies. Before we investigate the texture of the different kind of feeling language associated with *becoming*, let us further consider why the idea of an authentic separate self or 'true self' is so culturally ubiquitous and validated that it has achieved what Foucault would refer to as a 'truth status' (cf. Rose 1989, 1996). We will examine the cultural validation of the idea of separation central to the authentic self within one particular media site: reality television. Reality television is particularly interesting as it reveals the existence of both the *performing* and *true* self, and how these are coordinated. We can then explore in more detail how some of the themes and ideas about the communicating body as a realm of authenticity *and* as a site of manipulation are associated or related within popular culture.

THE BODY AND PERFORMANCE

If we take a reality television show such as *Big Brother* as a micropsychological laboratory to observe such processes, we can see how a conception of the *natural* body is brought together with a particular conception of the *social* body. Thus the body and its social environment are brought together in a particular relationship that relies upon the paradigm or concept of *social* influence. The conception of the *social* body assumes that bodily communications are a code that can be learnt, and are usually referred to as a culturally validated and recognized realm of interpersonal skills. The successful performance of this realm is linked to effective communication (Kristeva 1989). However, this realm can literally be confounded by the way the body can give itself away despite the person's attempts to manage their impressions. These slippages, or non-conscious expressions, are viewed within the paradigm of social influence as manifestations or disclosures of *authenticity*. This is a kind of *pure* self which is at odds with performance. This relies on a clear and distinct separation between the individual and the social, the biological and the cultural and the self and other. Indeed, studies of non-verbal communication assume that the unmasking of the bare human body is a form of communication that we share with animals. It is this realm that is viewed by many as most instinctual, primitive, innate and universal about what it means to be human. As we have already seen, this view of the naturalistic body is made intelligible through Darwinian evolutionism, and assumes that the realm of biology is that which is most fixed and invariable to social influence processes. It is an inviolate realm or territory, which, as we have seen, sets limits on how the social impacts or impinges on us.

We have seen so far that studies of the 'communicating body' within the psychological sciences and the tradition of microsociology tend to equate *authenticity* with a *naturalistic* body. Authenticity is revealed through gaps and slippages that are associated with *truth* and *nature*. The body is taken to be the site that can both disclose *nature* and also provide the vehicle for the management of social encounters through the successful presentation of bodily expressivity. We have seen the way that this story about the naturalistic body has a cultural value and is reproduced in many different ways throughout a variety of cultural sites. We have explored its taken for granted nature within reality television, but I am sure you can think of numerous other examples along similar lines. However, we have begun to explore a more embodied paradigm linked to the concept of *becoming* that would suggest a conception of the body that relies upon connectedness and mixing, rather than singularity and separation. This paradigm would not ask the question 'is it nature or culture?' (Despret 2004b: 35). This assumes that the body is the site where physical and social boundaries are drawn between self and other. Rather, we might instead look for other stories, what Despret terms 'versions' that exist in the background. This is precisely the strategy she undertakes in relation to the case of Hans the horse, which reveals a rather different conception of the 'communicating body'. We saw how this version is made possible from the gaps and anomalies that could not be connected up through a conception of communication that relied upon a separation between horse and human. I will now explore this conception further so that we can appreciate how it troubles the concepts of separation and singularity upon which the concept of social influence relies.

EMOTIONAL CONTAGION

I framed this chapter through a focus upon the extraordinary and phenomena that are difficult to explain within the paradigm of social influence. One such phenomenon is 'emotional contagion', which refers to the ways in which feelings can be passed between people with the result that their moods can shift and change. This phenomenon is recognized within the clinical literature exploring therapist's experiences of working with their clients (Hatfield, Cacioppo and Rapson 1994). Hatfield and Rapson worked together as therapists and comment on, 'how easy it is to catch the rhythms of our clients' feelings from moment to moment and, in consequence, how profoundly our moods can shift from hour to hour' (Hatfield *et al.* 1994: 1). They describe this experience as a kind of being 'in tune' (ibid.: 16) that creates a synchrony of feeling with those around you. They draw on the concept of *entraining* that was also central to Game's (2001) conception of *becoming* that we explored in the Introduction. This phenomenon suggests that people can be linked and connected physiologically and emotionally and can communicate this through an exchange of feeling of which they are not necessarily consciously aware. They chart the long history of this realm of affective exchange in clinical research, literature, the psychological and behavioural sciences and in events that have occurred within populations throughout history that involve the passing of mood, emotion and passion. One such example that they recount is cross-cultural evidence that documents various epidemics of laughter, depression, mania and seizures that have occurred in Singapore, Malaysia and Africa.

Although such 'contagions' have been documented throughout history there is plenty of evidence to suggest that affective exchange is a phenomenon that is part and parcel of everyday encounters. It is usual to focus in the literature on exceptional phenomena such as the mass waves of panic that followed Orson Welles's legendary radio broadcast of 'War of the Worlds' in the United States in the 1940s (Cantril 1940; Bourke 2005; Orr 2006). Many listeners actually believed that this was an unfolding reportage of the invasion of America by alien visitors, which caused 'mass hysteria'. However, I would like to turn to a recent study that documents the centrality of affective exchange in intimate relationships. A recent study carried out by Nick Powdthavee at the University of Warwick, presented at the Royal Economic Society's Annual Conference in Nottingham (21–3 March 2005), was commented upon in the print media. The stories were framed around a central question which the study purported to answer: Could your spouse's happiness determine your own happiness? The articles did not challenge the idea that happiness might be contagious and inhere between individuals; rather, the sensational aspects of the stories revolved around the findings which suggested that the only couples to benefit from such good feeling were married couples. What were clearly documented were the contagious aspects of happiness and how, for the married couple's, this buffeted them against the stresses and strains of losing a job, coping with illness and whether they owned their own property.

SELF-CONTAINMENT AND OTHERING

The psychological literature does not contain a unified or coherent explanation of these processes. Indeed, what marks the literature is its puzzling challenge to the idea of the individual being self-enclosed, clearly bounded and separate from others. Hatfield *et al.* (1994) liken emotional contagion to a form of magic that is little understood (within prevailing models) but has huge implications for public policy and for the practising knowledge of doctors, lawyers and therapists. As they recount, 'We may believe we guide ourselves through our daily treks, but a moment's reflection shows we neither proceed alone nor have as much control as we might have thought over others or our interactions with them' (Hatfield *et al.* 1994: 190).

The puzzling challenge of 'emotional contagion' to work on the body within social theory has been picked up by Teresa Brennan (2004) in her book *The Transmission of Affect*. Although this book is advanced reading, I would like to draw out some of the key concepts that she deploys in order to refigure the body as a body that is connected and not separate from others. She suggests that the huge field of documented instances of the transmission of affect are important as they break down the distinction between the individual and the social and the natural and the cultural. She argues that 'the transmission of affects means that we are not self-contained in terms of our energies' (Brennan 2004: 6). She turns the question of affective self-containment on its head. Rather than presume we are self-contained and separate from others, she directs the question to historical material that would suggest that there have been 'different, more permeable, ways of being' (Brennan 2004: 10). She looks rather at how we maintain an image of self-containment in the relationships that we develop with ourselves and others. One process that is part of the formation of a separate, singular body is one that relies on *Othering*. We encountered this process in Chapter 1 where we explored how within Darwinistic evolutionism the bodies of certain groups, such as women, colonial subjects, people with different sexualities and the working classes were considered inferior and degenerate in relation to a white, male, middle-class norm (cf. Blackman and Walkerdine 2001). Their bodies were Othered and viewed as the site of animality, primitivism and irrationality. We will also explore work on Othering in the context of the Body and Difference in Chapter 3.

Brennan argues that these Othering processes occur on a mundane level between individuals. Rather than recognize the permeability of boundaries and the transmission of affect we deny that affects are coming from the other, or deny that they are coming from us. In other words, we draw limits and boundaries around what we are willing to recognize, which often means that certain people are made to carry the affects of another. This failure to recognize means that we deny the emotional and affective connections that sustain our sense of subjectivity. There is a language for this denial of affective exchange that Brennan locates within Freudian psychoanalysis. The assumption that 'Othering' relies upon 'projection' was central to Freud's theories of the unconscious. This can create particular relational connections between people in which, as Brennan argues, 'the person projecting the judgement is freed from its depressing effects on him or herself. However, he or she is dependent on the other carrying that projected affect, just as the master depends on the slave . . . a kind of hook on which the other's negative affect can fix' (2004: 111).

However, the affective language that people tend towards is one that emphasizes separation, and tends to cover over or occlude the rather different kind of language that psychoanalysis, for example, makes possible. This has been considered by critical psychologists such as Paul Stenner (1993), who has explored how projection becomes hidden in the kinds of narratives and practices that people tend to deploy to understand the basis of jealousy. Thus, he tells us that commonsense practice assumes that jealousy is a property of the self-contained individual; there are 'jealous types' who cannot control their feelings in relation to another. Stenner troubles this view of jealousy as a property of an isolated mind and instead approaches it as a *subject position*. The concept of a subject position is one that has evolved from discursive work within critical psychology drawing on the weak and strong constructionist work that we explored in Chapter 1. Thus the notion of a 'jealous type' can be used to position somebody as unaware and unenlightened, or as being emotionally weak or insecure. Both of these strategic uses of a 'jealousy narrative' or story have particular implications for the person positioning and being positioned. Thus, in the example that Stenner develops of a couple, Jim and May, Jim positions himself as enlightened and progressive and May as fragile, unstable and weak. Because of this relational positioning he sees himself as having to walk on eggshells, therefore crediting himself with the power to hurt or protect May according to his actions. This relational positioning relies upon the concept of separation and also hides or covers over the kinds of projections or 'Othering mechanisms' that maintain this splitting. What is important to this formulation of separation is knowing where you end and the 'other' begins. As Brennan argues, 'the Western psyche is structured in such a way as to give a person the sense that their affects and feelings are their own, and that they are, energetically and emotionally contained' (2004: 25).

AFFECTIVE TRANSMISSION

Brennan (2004) suggests, however, that the evidence provided by the literature that points towards a model of connectedness and 'affective transmission' rather than separation and singularity is overwhelming. As we have seen, this is the starting point for work within an embodied paradigm and provides the basis for a range of contemporary work within social and feminist theory that is attempting to explain the mechanisms through which affects are passed between people. This work does not discount work within the biological and psychological sciences. As we have seen, this position of hostility and suspicion towards 'the biological' was a key marker of work that is described as social constructionist or as originating within a model of cultural inscription. Rather it revisits contemporary and earlier models of corporeality and attempts to rework them within an embodied paradigm. Brennan, for example, revisits work within human endocrinology (the study of the effects of hormones on behaviour and mood), and argues that this is one area that has not been approached through a model of 'affective transmission'. Thus, she pulls together exciting work within social theory that is bringing into play a rather different notion of the 'communicating body' with work in the biological sciences on 'chemical entrainment'. 'Chemical entrainment' is a concept that is used to refer to the subtle effects of hormones and pheromones that are communicated via smell and touch and that demonstrate that humans can affect and be affected at the level of the nervous system. As she argues, 'if olfactory communication turns a hormone into a pheromone and changes another's affects, does it also change their hormones in a way that (temporarily) changes their habitual affective disposition. Are such changes, in turn, communicated by additional pheromones? If such cycles can be shown to hold in groups, then the contagion of affects has been explained' (Brennan 2004: 72).

However, this is mere speculation at this point as studies of human endocrinology start from the position of 'affective self-containment' rather than connectedness and mixing. Brennan argues that one of the problems for social theory, and the view of the body that is indicative of this contemporary work on the body, is that science tends to work within the paradigm of social influence. Thus, she argues, science holds back from alternative *versions* (Despret 2004a and b) or views of the communicating body, and is thus constrained by what she describes as a 'foundational fantasy' (Despret 2004a: 73). However, one strategy of the growing body of work on becoming (Despret 2004a and b) is to look for marginal work within the physical and biological sciences that raise problems and conceptual difficulties for the paradigm and assumptions of social influence models. Brennan turns to work within psychoneuroendocrinology – a sub-branch of work on hormonal systems that produces anomalies for the fantasy of affective self-containment. Psychoneuroendocrinology is full of experimental studies that demonstrate that hormones can affect emotion and mood, and that they appear to be passed between individuals. She argues that this work demands attention from social theorists interested in corporeality and the materiality of the body, and offers a fertile starting point for new alliances between social theory and the biological and physical sciences. The aim of an alliance or dialogue between science and social theory is the goal of a diverse range of studies working within a more embodied paradigm. We will also cover similar work in Chapter 4 that works closely with studies in the medical and biological sciences, but approaches this from the perspective of the *lived body*.

THE CIVILIZED BODY

So far, throughout this chapter, we have seen the way in which a particular version of the communicating body, as one that is ideally separate, with clear boundaries between self and other, human and non-human, has been central to traditions within the biological and psychological sciences. It has also had a place in work within more sociological traditions. Indeed, many people argue that sociological work that turned to *social* relations as the key determinant of what it means to be human (within social constructionist traditions, for example) creates a range of different assumptions of separation that are equally problematic. This includes, as we saw in Chapter 1, separations between the biological and the social, structure and agency, passivity and activity and the body and the self. There is a tradition of work within sociology that turned to historical work on the body to explain how and why the idea of separation became so central to the way in which advanced liberal democracies tend to 'think the body'. Work on the emergence of the 'civilized body' has been given a central place within work on the sociology of the body that we explored in Chapter 1. The term 'the civilized body' refers to the ways in which the body in Western societies is 'highly individualised in that it is strongly demarcated from its social and natural environments' (Shilling 1993: 150).

The studies of the sociologist Norbert Elias (1994, 2000) have been drawn upon to illustrate the historical emergence of such a body, which Elias ties to a set of 'civilizing processes'. Elias employs the term 'civilization' rather differently to the ways used in the Darwinistic assumptions of civilization that we explored in Chapter 1. In Elias's work, 'civilization' is not an evaluative judgement used to demarcate and differentiate bodies that are considered inferior and more primitive from those that are considered rational and superior. Rather, he explores how different relationships to the body of the self and other were constituted so that 'body management' became the norm. 'Body management' refers to the ways in which bodily expression were increasingly to become matters of individual emotional control throughout industrialized societies. The norms of emotional control are such that individuals, Elias argues, are required to experience themselves as separate from others. Elias contrasts medieval societies with so-called civilized societies (that is, post-industrial) to demonstrate how bodies were considered as being, in contemporary terminology, more leaky and permeable, rather than as exhibiting and expressing forms of bodily and psychological distance. Elias's history of the kind of communicating body that is central to the paradigm of social influence offers a thoughtful challenge to some of the very many assumptions of this work. It is thus a very useful guide for readers who are interested in the historical antecedents of the problem of separation.

Similarly, both Despret (2004a and b) and Brennan (2004) turn to historical work that demonstrates how contingent the idea of 'affective self-containment' is to modern understandings of the body. Brennan reviews historical literature that suggests that the idea of affect flowing between individuals has a long history that is equalled by a very detailed and nuanced affective language. She argues that affects have to be split off both from conscious thought and from a sense of relationality for the self-contained individual to come into being. As she argues, we 'were once aware of the transmission of affect, but we have sealed against it' (Brennan 2004: 117). As we have seen, this is not simply a theoretical or conceptual problem for social theory and the kind of body that it might wish to make central to its analyses. This has very real consequences for the ways we relate to both ourselves and others, where 'the illusion of self-containment is purchased at the price of dumping negative affects on that other' (Brennan 2004: 119). One of the more ethical consequences that Brennan draws from her study is that individuals should be encouraged to think and reflect in more relational ways. This is not simply through cognitive reflection, which assumes a split between body and mind, but through the development of practices that work with 'the intelligence of the flesh' (Brennan 2004: 140). This is a rather different notion of the body that assumes that, rather than simply being inert mass, the body thinks and feels. We are going to explore this notion of the 'somatically felt body' in the next section as it has become central to related traditions in social theory exploring different versions of the communicating body.

THE SOMATICALLY FELT BODY

There is an emerging tradition of work within studies of the body exploring the potentialities of the body that cannot be contained through the concepts of cognition and the mind–body split which this assumes. The focus of much of this work is on what is occluded, hidden or swept away as insignificant by the privileging of cognition. One focus of this work is on forgotten thinkers or forgotten ideas in the work of thinkers that have taken up a place in social theory. Work is revisited and re-evaluated to explore what was taken up and what was firmly kept in the background. This strategic re-evaluation of what has been lost or forgotten is important in understanding how present presumptions of a mind–body dualism came into being. One example of this work is in a book written by a political theorist, Tamborinino (2002), which explores the writing of key philosophers who are central to political and social theory. These include the writings of Hannah Arendt and Frederick Nietzsche. Tamborinino argues that there are important yet silenced aspects of their work that present rather different formulations of the body. These formulations reveal what he terms the 'corporeality of thought' (Tamborinino 2002: 10), that is, a conception of thought that does not separate thinking from the body.

He suggests that in the writings of Nietzsche we can see a framing of the importance of affect to the body's potential to think and feel. The concept of *affect* is important to this work and refers to a realm of feeling that is not self-contained and separate but rather enhanced and produced through the relations between the self and other. Indeed, if we consider affect as a process that binds people together so much that they are not separate, then the idea of the self and other also needs to be rethought as a realm of *connectivity* rather than separation. Tamborinino draws on what he terms 'dissonant' or silenced aspects of Nietzsche's writings to bring to the foreground the concept of affect as a process that connects. Affect both inheres within people's bodies, and is thus felt, and also passes between them, meaning that we can affect and be affected. This attention to bodily feeling and affect is one that draws attention to those aspects of bodily feeling that are usually dismissed as involuntary and automatic. One such example that Tamborinino gives is the notion of 'gut feeling' or 'gut reaction' whereby one might have a particular sense or feeling for a person or situation without being able to explain why. Tamborinino suggests that gut feeling reveals a form of bodily intelligence that works on a level that cannot be explained by the concept of cognition. He suggests that this is because cognition has historically become associated with a separate realm of the mind capable of conscious reflection and deliberation. 'Gut feeling' discloses, Tamborinino suggests, the capacity of the body for a form of intelligent thinking that is *felt* rather than revealed through a verbalized language.

He also suggests that most of us already incorporate this form of intelligent bodily thinking into our judgements, revealing the importance of the concept of *attunement* in our encounters and interactions. Within this formulation of the 'intelligence of the flesh', reflection becomes a bodily activity arising from a body that is not separate from the mind and that connects the body to the other, human and non-human, with the result that one can affect and be affected. Thus, for Tamborinino, the 'turn' or 'return' to corporeality within social theory is also revealing aspects of our embodied existence that have been submerged and forgotten. The practice of resurrection and re-evaluation is becoming a marked tradition of work across critical psychology, sociology, cultural studies and feminist studies (Blackman 2007a and b, 2008a and b). The consequence of much of this work has been to reveal how our current conceptions of the body rely upon a silencing of central concepts that would point towards rather different bodily formulations. These are bodily formulations that challenge Cartesian dualism and foreground 'the richness of the affective and tactile–kinesthetic body'. This is what Sheets-Johnston refers to as the body of felt experience' (1992: 2). I will give you some examples of this work throughout the next section. Although it is advanced it is taking up a central place in 'thinking through the body' and thus is important to any engagement with the place of the body within social and cultural theory. The work is transdisciplinary and links and draws together different domains and realms of experience. We have already encountered the transdisciplinary nature of much of this work in our engagement with the writings of Despret (2004a and b) and Brennan (2004), who both bring together work at the 'hard' edges of the biological and psychological sciences with work on becoming within social theory. I am using the term 'hard' to denote theories that claim to be explaining human and psychological life through biological and physical processes. I am now going to expand this by giving you some examples from a book edited by Maxine Sheets-Johnston, *Giving the Body its Due*, that brings together work on the somatically felt body from anthropologists, psychologists, psychotherapists, artists and scholars interested in dance and Eastern psycho-spiritual traditions.

THE VITALIST BODY

This book explores what is occluded or silenced by Cartesian dualism in the context of medical science and related traditions. We have looked at how Cartesian dualism operates on the basis of a separation between mind and body and assumes that the body is 'a purely physical object' (Sheets-Johnston 1992: 2). We have examined how this conception of the naturalistic body ignores the idea of the body also being a feeling and thinking body. This approach to the body is captured by the concept of the somatically felt body or the *vitalist* body. The book contains many examples and instances of experiences that are difficult or impossible to explain from the perspective of Cartesian dualism. One such example that is referred to in different ways throughout the book is how to explain or understand 'the body that responds to placebos' (Sheets-Johnston 1992: 12). Placebos are inert pills, often sugar pills, that are given to subjects who are led to believe that they have certain active properties, in other words, that they will alleviate the problems and conditions with which they have presented. Placebos are used primarily in what are termed 'double-blind' trials. These are medical and scientific trials in which new drugs are being tested, and which require the existence of a control group to compare the results. The control group is often given such inert sugar pills in order to compare outcomes between the two groups. One of the most striking conclusions to have been reached by those studying the results of the trials is that subjects often get better by taking the placebos alone. (See Lakoff 2007 for a development of this view in relation to the concept of 'placebo responders'.)

In light of these unexpected findings, placebos have increasingly been tested against a control group who receive nothing with thought-provoking results. As Moerman asks, 'How does knowledge get transformed into physiological action? How does receiving a placebo pill heal ulcers? How does receiving an injection of sterile saline stop pain?' (1992: 81). All of these outcomes have been achieved by the ingestion of inert sugar pills. Moerman gives numerous examples of effects that have been achieved by the ingestion of placebos, including a study carried out that discovered that a placebo tablet for headaches had a more pronounced placebo effect if it was designed in such a way that it mimicked a leading brand. Other studies have found that injections or shots of a saline solution are often more effective than the ingestion of a pill. The placebo effect has even been exploited by pharmaceutical companies who have explored how the colour of certain tablets for the relief of depression makes a difference to the outcome. Thus, in a British study it was found that the tablet oxazypam was more effective in treating depression if it was dyed yellow (Shapira, McClelland, Griffiths and Newell 1970). One of the explanations often given for such effects is what is termed the 'mind over matter' argument. This reproduces Cartesian dualism by arguing that somehow the body is being influenced by the mind in ways that we do not really understand. The term that is often used within this tradition is 'psychosomatic', which recognizes that the body can be influenced by belief in curious ways (see Greco 1998). However, this approach still starts from the basis that the mind and body are separate and somehow interact.

Sheets-Johnston (1992: 147) suggests that the strangely 'curative powers of the body itself' point towards *connectivity* rather than *separation* of different entities or elements. She argues that Cartesian dualism although recognizing 'interaction' presumes that the body is primarily a 'de-animated' or 'physico-chemical' body (Sheets-Johnston 1992: 135). It is assumed (and indeed medical science is based on this assumption) that it can be separated from the mind and studied and approached as a functionally separated body made up of distinct systems such as the digestive, respiratory and nervous systems, for example. What guides medical practice, she suggests, is a concern with place, perhaps captured most tellingly by the question: *where* does it hurt? She calls for a more vitalist conception of the body, which, she argues, is already in the making, and indeed existed prior to the emergence of Western medical science (Foucault 1973). She resurrects theories that have now been discounted that worked with two concepts: *proportionality* and *relationality*. Proportion recognized that in order for a body to be healthy it needed to be 'in harmony'. Thus the ancient Greek theory of humoral medicine worked on the assumption that blood, phlegm and bile needed to exist in proportion to each other with excesses or restrictions of one part affecting the distribution of the other elements, causing disharmony. Curative regimes were based upon practices that aimed to redistribute the elements to eliminate overabundance or deficiency of particular elements. These practices did not simply work with the singular body of medical science. The other key concept central to these traditions was *relationality*. This concept recognized that the body could affect and be affected by others: human and non-human. As Sheets-Johnston argues, 'a body was in and of the world and as such was affected by other bodies, by the atmosphere, the seasons, the air, the water, the city and so on' (1992: 141).

THE NETWORKED BODY

Although medical science has largely lost the concepts of *proportionality* and *relationality*, she argues that anomalies like the placebo effect highlight the importance of concepts that were once present and may yet resurface in a new guise. She introduces the concept of the 'networking body' that does not assume separation but rather systems or elements that communicate with each other. We are back with a rather different conception of the communicating body that we have explored throughout this chapter. This is a communicating body that is characterized by connectivity rather than separation. This conception of the communicating body is central to many recent interventions made by feminists and cultural theorists who are seeking to refigure the kind of body that might be brought back into social theory. One example of this work is a recent book by the Australian feminist cultural theorist Elizabeth Wilson (2004) titled, *Psychosomatic: Feminism and the Neurological Body*. Although this is advanced reading it is guided by many of the revisions of the body that we have been exploring throughout this chapter.

Wilson's aim is to introduce an approach to corporeality or the materiality of the body that neither ignores biology, nor assumes that it is an inert phenomenon. What she does is to refuse the separation between the psychological and the biological brought into being through the idea of 'mind over matter' that we explored in relation to the term psychosomatic. She argues, rather, that the digestive, nervous and other so-called discrete systems can be demonstrated to be psychologically *attuned*. That is, they can communicate in such a way that the body is taken to be networked or to exhibit a 'relational complexity' (Wilson 2004: 20). She argues, therefore, that the 'placebo effect' is not simply 'in the mind' but demonstrates how other parts of the body have the 'capacity for psychological action' (ibid.: 34). As she argues, 'despite the large amount of clinical and anecdotal evidence that points to the gut's highly mobile, highly sensitive psychological nature, the psychodynamics of this part of the nervous system remain understudied' (ibid.: 33). Rather than dismissing the neurosciences she also aims to 'build a critically empathic alliance with neurology' (ibid.: 29).

THE FEELING BODY

This work thus exhibits many of the hallmarks of study that is emerging within contemporary body theory. One hallmark is the attempt to build bridges with the biological sciences, rather than dismissing them as *reductionist* and *essentialist*. As with the work of Teresa Brennan (2004), this paradigm of new work on the body seeks to explore what the biological as well as social sciences might look like if they started from the presumption of connection rather than separation. This, as we have seen, is a more *embodied* perspective that refuses or rejects the concept of social influence. This paradigm is producing some exciting work on the body that is developing an innovative language of affectivity – that is, the body's capacity to affect and be affected. One of the striking conclusions of this work is that Cartesian dualism covers over or silences the body's capacity for thought or what we might term 'the intelligence of the flesh'. Wilson's work is important as she is also attempting to revise what we might understand the psychological to be. She argues that the psychological is often used to refer to those processes that occur within the mind: that are cognitive. The psychological is seen to refer to those so-called higher processes that attempt to bring the body under control. We have explored this in relation to the concept of will and the idea of body-management introduced by the sociologist Norbert Elias (1994, 2000). However, her research demonstrates that the psychological is distributed throughout the body. As we have seen, the body is psychologically attuned.

Alongside this research there is also an emerging tradition of work that is calling for a more vitalist body to be brought into sociology and related disciplines. This work, which is being carried out by many of the sociologists who were part of the emergence of the sociology of the body in the 1980s, is guided by a renewed interest in affect, sensation and perception (see Fraser, Kember and Lury 2005; Lash 2006). In other words, a direction characterized by a concern and interest in a particular kind of *feeling* body. The difficulty posed by the 'inert body' was recognized as part of the problem that the sociology of the body had to contend with from its inception. As Lash argues, 'the body should possess some positive, libidinal driving force' (1991: 277). This is now a central concern and is producing a body of work that is developing new languages for talking about feeling and affectivity. This work demonstrates that the body which has been taken for granted across the sciences is one that has silenced and obscured more vitalist conceptions of the body that have existed, and indeed still exist at the margins of the medical, psychological and biological sciences. It challenges the idea that the mind and body are two distinct substances, and that thinking is a process carried out in separation from the body.

Johnstone (1992) argues that indeed the bodily nature of the self was obscured by Descartes because he refused to take on board the views of one of his female disciples, Princess Elizabeth of Bohemia. Descarte's views of dualism, he suggests, were based upon a fantasy that he could exist without a body. Descarte suffered with physical infirmities and produced his philosophical speculations on a misguided fantasy that he could overcome the body (Porter 2003). However, his female disciple constantly reminded him of those aspects of embodied existence that he chose to ignore. This included infirmity, as well as the effect of 'troubles, worries, and emotional turmoil on clear philosophical reflection' (Johnstone 1992: 22). Thus, certain aspects of existence were placed in the background including the 'tactile-kinesthetic or felt body' (Johnstone 1992: 28).

This is what was left behind and forms the silent legacy of Cartesian dualism.

Conclusion

Academic debate in the early twenty-first century has given rise to calls for a revision of the 'thinking self' that are introducing a conception of the body that is governed by a number of key concepts: *connectivity*, *proportionality*, *relationality*, *attunement*, *becoming*, *affective transmission* and *somatic feeling*. We have explored the emergence and meaning of these concepts in different ways throughout the present chapter. What this work emphasizes is the radical intersection of nature and culture, the biological and the social, the inside and outside and the human and non-human, to the extent that the idea of discrete entities interacting is beginning to lose its explanatory power. Rather, what we start with is an assumption of the permeability of boundaries and the inextricable connection of mind with body, human with non-human and biological with social. This work also introduces a 'non-cognitive' conception of embodied thought (Thrift 2000). That is, a tradition of study that is exploring the potency and centrality of non-conscious perception, of that which occurs below the threshold of conscious thought and deliberation. As Thrift maintains, many philosophers and sociologists have argued that 'much of human life is lived in a non-cognitive mode' (2000: 36). In Chapter 3 we will turn to work that has explored how social differences, such as gender, class, race and sexuality have become embodied, often in ways that become habitual and non-conscious. This chapter will focus upon the body and difference and will explore work across sociology, feminist studies, critical psychology and cultural studies that has extended the concept of cultural inscription to examine how raced, sexed, classed and gendered differences pass into our very being and becoming.

3 BODIES AND DIFFERENCE

The body and bodily dispositions carry the markers of social class

Skeggs *Formations of Class and Gender: Becoming Respectable*

INTRODUCTION

This chapter will focus upon a range of different studies across the disciplines of sociology, feminist studies, critical psychology and cultural studies that have taken analyses of the body as central in understanding how social differences, such as race, class, sexuality and gender, pass into our very being and becoming. I have started the chapter with a quotation from the British sociologist Beverley Skeggs, who has produced a volume of work exploring how class is lived, embodied and enacted by those positioned as Other to middle-class rationality (1997, 2004). The concept of Otherness is crucially important for understanding the direction of this work, and, as we saw in Chapter 2, is integral to how a concept of the separate, self-contained subject is brought into being. In Chapter 2 we explored some of the writings of the late feminist writer, Teresa Brennan (2004), who argued that in order for subjects to take up a culturally valued position where they see themselves as 'affectively self-contained' they are encouraged to participate in what she terms 'Othering processes' or mechanisms. These are processes in which it is denied that affects are being passed between subjects. She argues that rather than recognize the permeability of boundaries and the transmission of affect we deny that affects are coming from the other, or deny that they are coming from us. In other words, we draw limits and boundaries around what we are willing to recognize, which often means that certain people are made to carry the affects of another. This failure to recognize means that we deny the emotional and affective connections that sustain our sense of subjectivity. These Othering processes do not simply exist on a mundane interactional level between social subjects. They are also enacted and reproduced across a range of material and social practices that position actual bodies in relation to regulatory ideals. This positioning produces certain bodies as inferior, lacking, dangerous, deficient and abnormal. This is the realm of the Other as it is embedded and enacted across a range of different sites including cultural and social practices, media cultures and legal and governmental practices (see Blackman and Walkerdine 2001).

One of the key concepts for exploring the relationship between social processes (of Othering) and embodied subjectivity is the concept of cultural inscription, which was introduced in Chapter 1. As this term suggests, what is important is how social or cultural processes inscribe or speak through individuals. These processes are so manifested in the thoughts, actions, bodily dispositions and habits of subjects that they appear natural and automatic. The body, within accounts of cultural inscription, is important for understanding the workings of ideology and power and for how social processes are seen to take hold. However, as we saw throughout Chapters 1 and 2, what is brought into the analysis, albeit in an under-theorized way, is a view of the body as a malleable entity that cannot speak back. The body, although central, is re-viewed as inert mass or matter that is brought under the control and disciplining of regulatory practices. It is what we might term an un-*thought* and certainly an un-*felt* body. Most of the studies that we will cover in this chapter start from a similar position to that which was expounded by early sociologists of the body and formed the focus of Chapter 1, that is, the paradox and problem of recognizing that the body is not pre-formed and pre-social whilst not wanting to view the body as socially determined in a way that removes from it any sense of agency or affectivity. There are a number of key concepts introduced in this tradition that attempt to

Case Study

If we take a media text such as the British make-over show *What Not to Wear*, actual viewers are invited to understand working-class female subjectivity through particular codes and structures that produce working-class female subjectivity as lacking, deficient and pathological. These are positions constructed within the text itself with which viewers are encouraged to identify. Angela McRobbie (2005) has argued that a media text like *What Not to Wear* has a particular cultural effect that she relates to Bourdieu's concept of *symbolic violence*. The concept of symbolic violence captures how discrimination and forms of oppression do not have solely to be enacted in a physical realm. The notion of violence being manifest in a symbolic realm points towards the central role that processes of signification play in positioning certain bodies as abject and deficient — in need of transformation bridge both biological and discourse or social determinism: what Fuss (1990) terms a more sophisticated essentialism. These concepts will be presented by starting with the work of Beverley Skeggs who has drawn on the work of Foucault and the French sociologist Pierre Bourdieu to 'think through' this problem. We will then proceed to related work that draws on phenomenology and more psychoanalytically inspired concepts to introduce a sense of bodily affectivity that relates to some of the issues that we explored in Chapter 2. The question then becomes one of how to think the relationship between matter and affectivity, and this will form the focus of the latter part of the chapter. We will consider critical approaches to subjectivity that have placed understandings of bodily affectivity at the centre of understandings of the relationship between being, becoming and difference; that is, how bodies come 'to mark and be marked with (in)equalities' (Schiebinger 2000: 8).

BODILY MARKERS OF RESPECTABILITY

Respectability is one of the most ubiquitous signifiers of class. It informs how we speak, who we speak to, how we classify others, what we study and how we know who we are (or are not). Respectability would not be of concern here, if the working classes had not consistently been classified as dangerous, polluting, threatening, revolutionary, pathological and without respect. It would not be something to desire, to prove and to achieve if it had not been seen to be a property of 'others', those who were valued and legitimated. (Skeggs 1997: 1)

or a 'make-over'. McRobbie argues that the two presenters of the show 'Trinny and Susannah' are both middle-class professionals who are credited with having a superior knowledge about what one ought and ought not to wear. They are *cultural intermediaries* (arbiters of what counts as good taste) who bring to bear forms of knowledge or *cultural capital* that they have acquired through their positioning as experts within a specific social field – in this case fashion and styling. Through their position as middle-class professional stylists they generate, carry and display bodily dispositions expressed through accent, clothes and bodily comportment that signal their superior social distinctiveness. They are respectable, trusted experts who are there to judge, evaluate and transform the willing participants, who are often presented as in dire need of such transformation.

The focus of Skeggs' studies (1997, 2004) is on how classed differences pass into our very being and becoming so that they appear as if they are natural and inevitable bodily markers. As Skeggs argues, one of the most culturally salient and ubiquitous signifiers of class in Western advanced democracies is respectability. This signifier, Skeggs argues, has taken on the status of a *social distinction*. The concept of social distinction is taken from the work of the French sociologist Pierre Bourdieu (1984), who explored in his seminal study *Distinctions: A Social Critique of the Judgement of Taste* how the French class system was literally written into the bodies and minds of the middle and working classes engendering certain entitlements, statuses, constraints and restrictions. Bourdieu conceptualized these asymmetrical and differential classed distinctions through the concept of *capital*. Class is not merely considered a structural position that engenders differential access to income, wealth and monetary assets but also as organized sets of relationships that distribute other kinds of capital throughout populations, including what Bourdieu termed *social, symbolic* and *cultural* capital. Social, symbolic and cultural forms of capital are acquired, learnt and formed through social practices such as education and schooling and can be used as cultural resources to position and be positioned within social relationships. They are marks of status and social differentiation and manifest through *bodily dispositions*. Bodily dispositions are ways of talking, walking, eating and conducting oneself, for example, that are judged, legitimated and recognized through hierarchical distinctions made between the superior and inferior, between those who are considered to have good taste, and those whose tastes might be considered vulgar and cheap (Harbord 2002). Bourdieu suggested that these distinctions are primarily classed distinctions that produce bourgeois middle-class taste as that which is viewed as normative, natural and as expressing the quality of superior distinctiveness. In other words certain forms of cultural capital are legitimated and these forms of articulation create *subject positions* that actual subjects have to negotiate, inhabit and embody in different ways. The concept of subject position is taken from the work of Michel Foucault that we explored in Chapter 1. Subject positions could be thought of as invitations to see oneself or others as a particular kind of subject.

CORPOREAL CAPITAL

Skeggs (1997) identifies how the appearance of working-class women is both a site of reproduction in relation to middle-class norms and also a site through which they might resist such pronouncements. One of the salient findings to emerge from Skeggs' study of white, female, working-class caring professionals was how they were very aware that clothing provided access to particular forms of cultural capital. These forms of cultural capital were linked to specific competences (knowledges) about what is deemed appropriate and inappropriate in terms of respectable dress and clothing. Although many of the women aspired to respectability and did not wish to be recognized as working class, they also expressed doubts, anxieties and insecurities linked to their very real perceived lack of status and entitlement. Their investments in middle-class respectable femininity were structured by a desire to refuse the Othered working-class feminine subject position that has been constructed historically as vulgar and lacking in discipline. Thus, bodily appearance, its fashion and stylization, is a salient form of *corporeal capital* for the women in this study who come from a poor northern English town and have little opportunity for social mobility and transformation. So it is that their experiences of inequality are embodied through a range of dispositions expressed through the surface markers of the body such as fashion and adornment, as well as more embedded bodily traits and dispositions. These include weight, height, gait, accent, comportment and an accompanying range of affective responses such as shame, guilt, anxiety, insecurity and doubt linked to their awareness of negative class distinctions. As Skeggs argues, 'the body is the most indisputable materialization of class tastes (1997: 82).

We can see from the example above the way in which concepts derived from Bourdieu's sociological approach to the body can be used to enable a link to be explored between the social articulation of Otherness, and how this is incorporated, enacted and embodied by actual subjects attempting to negotiate such differential positions. We can see the way in which different forms of capital have an 'exchange value' (Crossley 2001) and enable or constrain certain possibilities for action and entitlement. As Crossley argues, 'social practices are incorporated within the body only then to be reproduced by embodied activity' (2001.: 92). These are not 'innate materialities' of the body, but matter as it is materialized through socially mediated processes of regulation and disciplining. The crux of Bourdieu's approach to the body that connects it with work on cultural inscription that we explored in Chapter 1 is the focus on *reproduction* and *regulation* of bodily norms rather than their resistance or transformation. There is a sense as with much of the early work on the sociology of the body that there is a lack of attention to a sense of bodily agency which might enable the body to speak back or refuse those positions it is invited to inhabit. Therefore, many of the concepts are rather deterministic and are supplemented, as in the work of Skeggs (1997, 2004), with a consideration of a realm of bodily affectivity that also signals the lack of a tight fit or homology between subject positions and subjectivity. We will explore this realm of bodily affectivity in the next section by considering more phenomenological approaches to bodily affectivity as a way of further understanding how bodies are moulded and transformed through embodied practices. What we will see is that an understanding or awareness of classed distinctions, for example, is not only enacted through bodily dispositions but through a realm of *feeling* or bodily affectivity that cannot necessarily be known or articulated very easily (Charlesworth 2000).

FEELINGS AND BODILY DISPOSITIONS

> We find persons or things becoming or unbecoming, beautiful or ugly, and this affects our responses and relations, creating fields of force, or a dimension to human existence that is felt through affinity, distance or repulsion, whose processes lie deep in the socialized body; a kind of bodily kinetic sensitivity, of unerring logic, that has grave consequences for individuals whose world and being fall towards the negative pole of social valuation. (Charlesworth, 2000: 18)

Charlesworth's study is also of a poor, working-class northern English community (Rotherham, South Yorkshire) and the ways in which its members attempt to articulate or indeed are often unable to articulate their experience of dispossession. What Charlesworth point towards, which is significant for our consideration of bodies and difference, is the importance of introducing a more kinaesthetic sense of bodily affectivity for understanding socially mediated bodily dispositions. We looked at the concept of the feeling body or the kinaesthetic body in Chapter 2. In this chapter we will review studies that have turned to this realm to explore 'the opposition of distinction and stigma' (Charlesworth 2000: 69) in the lives of those who are positioned as Other. In other words, what it feels like to be marked and fixed as a particular kind of subject (Skeggs 2004). This is a focus not just on what people *do*, how they enact classed, raced, gendered or sexed distinctions, for example, but how they *feel* and what these feelings might reveal about their embodied subjectivities. Charlesworth draws on the work of the French phenomenological philosopher Merleau Ponty (1962) to argue that when we perceive the world our awareness is lived and felt, not simply represented in a cognitive, visual register. We usually think of perception as a process that occurs primarily through a visual register where our observations and understandings are either accurate or distorted through some kind of cognitive bias. Gibson (1979) refers to this view of perception as assuming that the human subject perceives in much the same way as the fixed aperture of a camera, rather than with a more holistically defined 'kinaesthetic lived-bodily incorporation of the sense of the world' (Charlesworth 2000: 64). This sense of the world is registered in 'affective states or sensibilities that are social in the sense that they are prior to a particular individual's feelings and govern the range of feelings available' (ibid.).

Feelings are defined as 'somatic manifestations' (Charlesworth 2000: 77) of a person's orientation to themselves and others. These feelings are often unspoken and even impossible to articulate, and therefore are not written upon the body in the way that the bodily dispositions to which Bourdieu refers might be. They might be *felt* in the form of anxieties, pessimisms, feelings of inertia, constraint and worthlessness. They might be disclosed in what Charlesworth refers to as forms of affective comportment (2000: 104); these might be ways of coping that enable a person to get through the day. They are rarely reflected upon becoming part of the habitual responses that are learnt and enacted and that become part of the 'inherited background' (ibid.: 108). This is the background (of social difference and its cultural inscription) that Charlesworth suggests is disclosed at an affective level and therefore is very difficult to see and verbalize. We explored in Chapter 2 how there is a recent trend towards work on the body and affectivity across the humanities that is suggesting that affects can be passed between people (Brennan 2004). This critique of the self-enclosed, clearly bounded individual examines how the borders and boundaries between subjects are porous and permeable, meaning that the limits of the body are not defined by the skin for example. Charlesworth provides an interesting take on this notion of 'affective exchange' by exploring how 'inherited backgrounds' can be passed between people through silence or condensed words like 'crap'. This work points towards a more gestural form of communication that is enacted through minimal forms of bodily communication that do not respond to universal codes or patterns. Rather, they are situated in a shared experience and understanding of what it feels like to be positioned and have to negotiate a cultural realm of social difference and its articulation. This is what Charlesworth terms a 'felt relation to the world' (2000: 124). 'The silences and ellipses in people's speech are their implicit, unknowing recognition of the background; those moments when the unsaid, shared, unspoken, passes between people, manifesting in knowing silences and appropriate gestures' (Charlesworth 2000: 113).

BODILY AFFECTIVITY

Charlesworth argues, therefore, that it is centrally important to consider a realm of bodily affectivity if one wants to understand and analyse the cultural inscription and incorporation of social difference. He draws his conclusions from the series of interviews he made with working-class men and women living in Rotherham. One of the key findings of his study that relates to the discussion so far in this chapter is that although the interviewees were aware of their social positioning as Other and the stigma and inequalities this had generated, their interviews were marked by silence and being unable to speak. When they were able to speak of the pain of their lives this was often communicated through a kind of joviality, which, as Charlesworth suggests, appeared to be a defence against the sadness and shame they felt. Charlesworth turns to concepts from Gestalt psychology, as well as phenomenology, to make sense of these *felt* forms of bodily affectivity that are literally felt and expressed in and through the body. Phenomenology is a tradition of philosophical thought that works with the concept of a *sensient body*. The mind is not separated from the body as in Cartesian traditions but rather the body is a thinking body that perceives its environment through lived, felt forms of activity in which the mind and body are viewed as integrated processes. It is not that the mind perceives (cognition) and that this is translated into forms of bodily activity; rather, perception is a thoroughly embodied experience and activity that cannot be reduced and abstracted into component parts.

Hwa Yol Yung (1996) suggests that phenomenology is a more traditionally feminine way of understanding bodily experience based more on touch and sensation than masculine forms of rationalist thinking. Rather than being focused upon what we might term the 'mind's eye' (rationality), its focus is with the 'body's touch' captured by the idea of being hugged, embraced and caressed. The body is a *thinking body* that enables it to be attuned to certain significances in the environment, a sensitivity that is embodied in the notion of 'tactile intelligence' (Charlesworth 2000: 194). As

Case Study

One contemporary popular media fantasy that has been embraced by children and adults alike are the stories of Harry Potter. The extraordinary global success and appeal of these stories made J. K. Rowling the first children's novelist to become a multi-millionaire and reputedly the richest literary writer in history. It is claimed that they have sold over 300 million copies worldwide. The stories have been adapted for cinema by Warner Bros. Studios and filmed versions include *Harry Potter and the Sorcerer's Stone* (2001), *Harry Potter and the Chamber of Secrets* (2002), *Harry Potter and the Prisoner of Azkaban* (2004), *Harry Potter and the Goblet of Fire* (2005) and *Harry Potter and the Order of the Phoenix* (2007). The films are all in the top twenty-five of the highest grossing films alongside such classics as *Star Wars* and the

Lord of the Rings trilogy. The films and novels are described as part of a fantasy genre which places the lead protagonist Harry Potter in a fight against good and evil. The narrative structure of the collective tales are rather similar to those that Walkerdine identified within young girls' comic books such as *Bunty* and *Judy*. Harry Potter is an orphan who is raised by cruel foster parents. After stoically enduring this misery he discovers that he has magical powers and is in fact a wizard. He is invited on his eleventh birthday to study at Hogwarts, a school for wizards. Harry Potter learns to use his magic in an ethical, morally responsible, kind, courageous and selfless way that enables him to overcome a variety of emotional and social obstacles. These are hindrances and problems that are depicted as utterly normal and mundane and that would relate

Charlesworth suggests, the importance of bodily affectivity for understanding the relationship between matter and social difference is precisely in its call for a 'move beyond the model of the conscious-subject linguistically constituting experience' (2000: 201). This question of the non-conscious or unconscious has been central to feminist debates on the body and embodiment, and we will begin to explore these issues later in the chapter.

EMBODIMENT AND MEDIA CONSUMPTION

One of the most successful Hollywood films to refer to working-class male practices of the self is *Rocky*, staring Sylvester Stallone. Following its enormous success *Rocky* has been joined by *Rocky 2, 3, 4, 5* and the latest addition *Rocky Balboa* (2006), which all focus upon the make-over of the protagonist Rocky, through boxing, into a successful working-class hero. The rags-to-riches tale is structured around the significance of the signifier 'fighting' in the life of a young working-class man who is defined through his body (brawn) rather than the mind (brain) of the middle classes. His only option is to fight rather than engage in back-breaking forms of

to the worlds and realities of many of those viewers and readers of the texts.

Harry is contrasted in the books and films with other characters who also have magical powers but use these to their own ends in the pursuit of evil and death rather than for good and love. One of the most important themes to endure across the narratives is that what is important are the choices one makes about how to use one's magical powers and abilities. Harry makes the right choices (characterized by friendship, kindness and love), and these choices enable him to recognize and challenge the forms of discrimination and oppression that exist in the world of Hogwarts, which include differentiation between wizards according to their parental ancestry with pure-bloods, half-bloods and Muggle-born (those born to non-magical parents).

In relation to these forms of social distinction, one of the enduring themes of the tales are that prejudices can be overcome by practising love and kindness, which are epitomized through the heroic magical actions of Harry Potter and his close friends and companions Ron Weasley and Hermione Granger. In this sense the films mobilize and tap into the difficult feelings and transitions that accompany adolescence and offer resolutions that present normative forms of affective comportment such as being kind, courageous, selfless, loving and morally responsible in order to overcome prejudice and evil. It presents the normative subject as one who is able to overcome prejudice and discrimination through their own loving actions, exercising choice through paying close attention to the consequences of their (magical) actions for others.

manual labour in order to survive. His turn to boxing condenses these coping strategies into a practice that enables his transformation and self-improvement. Valerie Walkerdine (1990) has talked about the significance of this media fantasy in the lives of working-class consumers, who historically, within media studies, have been considered stupid, taken in hook, line and sinker by the media moguls of Hollywood (Blackman and Walkerdine 2001). *Rocky* could be considered a sexist, violent and oppressive media fantasy that reproduces a stereotype of the working-class male as being defined by his body, his perhaps threatening physicality. However, not to go beyond this superficial reading misses a crucial element of what we might define as a more embodied approach to media consumption. That is, that what we might term 'populist media fantasies', might be seductive or potent because they resonate with or channel desires that are already present in the lives of oppressed peoples. Walkerdine suggests that the signifier 'fighting' has a resonance in the lives of working-class people who, in the context of advanced democracies, are positioned in a contradictory and ambivalent fashion. On the one hand they are invited to see themselves as agents of change responsible for their own transformation through hard work, effort and struggle. However, in order to participate in practices of social mobility they are invited, on the other hand, to see working-class culture as a site of shame and vulgarity, a place from which one would definitely want to escape. Thus the signifier 'fighting' does not simply reproduce a violent, macho fantasy of escape but connects with the very real struggles and dilemmas that such a transformation might entail.

Walkerdine's approach to media consumption, and its relationship to the cultural inscription and enactment of social difference, also introduces bodily affectivity as an important part of processes of subject formation. Walkerdine (1990) uses the concept of desire to refer to a realm of affect and emotion that is produced through the individual's attempt to subjectively live and 'hang together' the different ways in which they are positioned as a subject. This affective realm is a felt orientation to the world that might be verbalized and expressed through recognized emotional vocabularies such as guilt, shame, fear, humiliation and longing. It might also manifest through a realm of feeling that is usually recognized through a more psychopathological vocabulary, such as depression or anxiety or through body practices such as anorexia or bulimia. Walkerdine uses the concept of desire to refer explicitly to an unconscious realm that is not easily verbalized. This is not a realm of repressed biological drives or instincts that would reproduce an essentialist or naturalistic body (see Chapter 1). Rather, this is a realm of suppressed and repressed desires (to include a complex realm of bodily affectivity) that are produced through our positioning in relation to a range of social and cultural practices. Certain desires have to be split off, repressed or projected outwards in order to construct a coherent and unified sense of subjectivity. Walkerdine argues that in order to understand media consumption one needs to explore how this affective territory might be recognized, channelled and resolved through particular media fantasies that work in a world that is not simply conscious and cognitive. This is the realm of the unconscious or non-conscious that is increasingly being recognized as an important part of body theory. We will turn to it explicitly in the next section.

The following example, taken from the book *Schoolgirl Fictions* (Walkerdine 1990), develops a method for thinking about the potency and seductiveness of certain populist forms of media consumption. The analysis focuses upon examples of such consumption common amongst young girls in the 1970s and 1980s. These were the British comic books *Bunty* and *Judy*, which included stories that repetitively reproduced a particular narrative tale or structure in a comic format. The stories revolved around distinctions drawn between 'good girls' and 'bad girls' whose characters were defined through their forms of 'affective comportment' (ways of coping). The good girls were those who were selfless, always kind, understanding and helpful in the face of often extreme adversity and trauma including being brought up in an orphanage, being abused by wicked foster parents or befalling an awful accident that would take them out of their familiar situation. The 'bad girls' in the narrative were always those who expressed anger, envy, greed or other dispositions that Walkerdine suggests reveal the affective responses that are actively dis-encouraged in the stories as normative forms of coping. The 'good girls' were always constructed as happy, hopeful, kind and courageous, and were rewarded in the resolution of the narrative by escape or transformation of the situation. Walkerdine suggests that certain media fantasies offer a 'working through' of the very real conflicts and dilemmas of 'growing up girl' in a patriarchal culture in the 1970s and 1980s. This was a culture which many feminist scholars have suggested constructed normative femininity as that which was viewed as lacking and inferior in relation to masculinity, and that could be escaped or transformed through a romantic heterosexual relationship. This cultural injunction and longing is epitomized in the saying 'One Day My Prince Will Come' (Walkerdine 1990; McRobbie 2000). Although this example might seem somewhat dated now, it does give us a way of thinking about the role media fantasies might play in recognizing, identifying, channelling and resolving an affective realm that we bring to media texts and that is manifested through a felt, bodily orientation to the world.

THE BODY AND INDIVIDUALIZATION

In Chapter 2 we explored how in Western advanced democracies one of the key concepts for defining the body that came into being across the biological and psychological sciences at the turn of the twentieth century was separateness rather than connectiveness. This is what the British sociologist Nikolas Rose refers to as the rise of the 'fiction of the autonomous self', which he explicitly links to the emergence of the psychological sciences (1989, 1996). Where bodies were viewed as permeable and porous, captured by the interest and even orchestration of mass forms of emotional contagion by historical dictators such as Hitler (see Chapter 2), the focus shifted across the human sciences to the investigation of the will. The will was viewed as a form of psychological control that would allow subjects to dominate their bodies, thus fortifying them against contagion, suggestion and social influence (see Chapter 2). This introduced a contrast between those who were able to resist social influence (suggestion), through exercising their rational minds and those who were viewed as overly or easily suggestible, defined by their bodies rather than the rationality of the middle classes. We have seen the way in which this refigured the mind–body dualism through mapping it on to historical distinctions that had already been established which constituted certain peoples as lower and inferior on an evolutionary scale. This is one aspect of the Othering of bodies in relation to the establishment of the fiction of autonomy that we have investigated in different ways throughout the chapters so far.

In contemporary advanced democracies, sociologists such as Nikolas Rose (ibid.), Anthony Giddens (1991) and Ulrich Beck (Beck and Beck-Gernsheim 2000) have argued, these processes have mutated into what is characterized as a process of *individualization*. Individualization characterizes the elevation of particular ways of thinking about individual choice and responsibility which have increasingly been proffered as resolutions to success, satisfaction and happiness in the workplace, re-lationships, health and well-being and so forth. Normative resolutions present part-icular forms of 'affective comportment' (coping) as the way out of misery, suffering and so forth. The practices that the individual is invited to participate in rely upon the individual being able to exercise choice and responsibility. The individual is required to see their life as the outcome of the choices they have made and to make these choices responsibly. Thus failure cannot register as bad luck or misfortune but rather as personal failure or inadequacy. In some ways we can see how a media text like *Harry Potter* mobilizes the difficulties of living this fiction of autonomy, as well as recognizing the injustices and Othering that it can entail. In this sense, attention to a realm of bodily affectivity might be a pertinent way of thinking about the success of certain populist media fantasies that recognize the potent ability of films to mobilize such a realm. As Vivian Sobchack (2004) argues in *Carnal Thoughts: Embodiment and Moving Image Culture*, being able to address the experience of being 'touched' by movies invites a more embodied approach to media consumption. These are what she terms the 'somatic' aspects of cinema going that bring the body more centrally into the frame.

We have dealt explicitly with the relationship between cultural inscription, the social articulation of difference and the question of how such a realm might be embodied and enacted at the level of individual, subjective experiences of bodily affectivity. Social and material practices, such as the media and screen cultures like film, play an important intervening role in circulating fantasies that have the potential to connect with this already existing realm, and to shape it in ways that relate to more normative and culturally and governmentally sanctioned practices of everyday life. This is not about the all-powerful media crushing the bodies of mass consumers and viewers, but, rather, certain media fantasies working with and alongside certain forms of *felt, bodily* orientation to the world which might be shaped and channelled in specific normative ways. This section has also pointed towards work that suggests that this realm and our experiences of it are not always easily available for verbalization and articulation. It is a realm that exists in the background, although exercising important directions on how we exist in the world. The next section will explicitly focus upon arguments, mainly made by feminist scholars, that have turned to psychoanalysis in different ways to theorize bodily forms of feminine subjectivity. As we will see, this area of *feminine becoming* in relation to the body and embodiment is not unified or coherent. There are many arguments and counter-arguments that exist in the field. The focus will be on defining some of the key concepts to enable you to engage with the discussion in a more easily assessable manner.

THE POLITICS OF FEMALE BODIES

The call for a return to the biological roots of the body reiterates the rejections of social constructionism which is crucial to feminist theory in the third millennium.

Braidotti, *Metamorphoses: Towards a Materialist Theory of Becoming*

It has been argued that the body is at the centre of feminist theory (Conboy, Medina and Stanbury 1997). Indeed, early sociologists of the body, whose work we explored in Chapter 1, turned to many aspects of feminist theorizing in order to bring the body more centrally into sociological thinking (Turner 1984; Shilling, 1993). As Howson (2005: 3) cogently illustrates, there is a very close link between body concepts in sociology and 'concepts of gender within sociology and feminism'. The feminist work that we will consider in this section privileges the body as being key to understanding the relationship between cultural inscription and women's lived bodily experiences. However, there are many tensions surrounding this work which link to some of the tensions that we have explored so far in this book. We will start by considering the politics of the female body and some of the paradigms within feminist theorizing that have made the body central to understanding cultural inscription. We will then be ready to explore some of the different ways in which the concept of *feminine becoming* has been mobilized in feminist theory to address the question of how social and ideological processes take hold.

The tensions and the different ways in which the concept of feminine becoming is mobilized relate to issues that we have already explored in Chapter 1 and Chapter 2. These include the debate between essentialist (naturalistic) and constructionist approaches to the body (see Chapter 1), the relationship between regulation and bodily agency, the relationship between the materiality of the body (matter) and cultural inscription and the relationship between normalization, Otherness and embodied subjectivity. We have seen the way in which the social articulation of difference 'functions by exclusion and disqualification' (Braidotti 1997). Social differences such as race, class, gender and sexuality, for example, are constituted through distinctions made between the normal and that which is taken to be Other to the norm. As we have seen, the Other is that which is usually constituted as inferior, lacking, deviant or deficient in some way. The norm is that which passes as common-sense and natural, usually remaining unquestioned, invisible, silenced and unmarked (see Dyer 1997 for a discussion of whiteness). The question in relation to gender and sexuality that has interested feminist theorists is how these cultural injunctions and subject positions might be literally written into the flesh of the body. This is usually framed as the problem of *materiality* in which the physiology and biology of bodily processes are reintegrated with approaches that stress the disciplining and regulation of bodies through social and discursive practices. In other words, the question of how to think the relationship between the social marking of bodies as Other, for example, and embodied subjectivity.

THE CORPOREAL TURN

Howson (2005) examines what has been constituted as the *corporeal turn* within feminism over the previous two decades during which time a more explicit focus on the body has taken place. The corporeal turn is largely a response and reaction to the predominance of a paradigm of social constructionist approaches to the body within feminism. We have already spent some time in Chapter 1 exploring social constructionist approaches to the body, which as we saw have tended to reduce the body to a discursive or textual effect. Although the body is viewed as a malleable and unfinished phenomenon, priority is given to the ways in which the body is sculpted and moulded through regulatory practices. This in effect marginalizes the materiality of the body and performs what many have argued to be a more sophisticated form of essentialism known as social or discourse determinism (Fuss 1990; Riley 1983). Howson argues in a similar vein that the focus within the paradigm of social constructionism on 'language, discourse and text' reframes the body as 'text rather than matter' (2005: 9). Although we have already examined the problems of both naturalistic (essentialist) and social constructionist approaches to the body within early debates that formed what came to be known as the 'sociology of the body' (see Chapter 1), we will further explore them in the context of feminist debates about sex and gender. These debates form part of the backdrop to the emergence of *corporeal feminism* and the concepts that have been developed to address the problem of materiality.

Feminism, arguably, has always expressed a particular politics of the body largely focusing on the subordination, marginalization and oppression of female bodies through sites such as (reproductive) medicine (Martin 1987; Duden 1991; Oakley 1984; Young 1990), pornography (Cornell 2000), advertizing (Gill 2006), popular culture (McRobbie 2005; Walkerdine 1997), cosmetic surgery (Balsamo 1996; Bordo 1993; Davis 1997), cyberspace (Braidotti 1996; Haraway 1991; Kember 2003) and the life and biological sciences (Hayles 1999; Haraway 1996; Kember 2003). One of the key debates that has structured feminism from its inception is how to approach and understand the status of sex and gender. Increasingly in the 1970s and 1980s second-wave feminism framed gender as a socially constructed category. Gender was not considered an invariable biological category but rather was viewed as a social concept that positioned women differentially in relation to men. This constructionist approach to gender is often linked back to the arguments made about femininity by Simone de Beauvoir in her seminal book *The Second Sex* (1953). Barbara Brook (1999: 11) reproduces one of the most quoted parts of the book, which has become foundational to constructionist approaches to gender, that 'One is not born, but rather becomes, a woman' (de Beauvoir 1953: 295). As Brooks goes on to argue:

> In all the various permutations of social constructionist theories there is at some level a basic distinction made between the material body and its social/cultural representations. So the body is seen as a kind of natural biologically sexed object that pre-exists but is affected by the workings of culture or, as some writers term it, a *tabula rasa:* a blank surface ready to be inscribed. It is this separation of body and culture that defines the sex–gender division. It rests on the belief that, whilst there are certain natural attributes of the body which cannot be changed (or not without radical surgery), the gendered cultural meanings circulating around and variously inscribing the body can be changed. And, of course, that process of changing cultural meanings to relieve the inequalities of women, in this definition, comprises the feminist project. (1999: 11)

The problem with the distinction between sex and gender as it is articulated in this way is that the body is relegated, denied or assumed to exist as a 'biological, natural, sexed body' (Brooks 1999: 12). It is culture that becomes the site of change and mutability, and the body located as the site of fixity and invariability. In this respect sex remains an essentialist category (usually aligned to the workings of hormones, anatomy and physiology) with gender overlaid as a constructive shaping force. It is this distinction between sex and gender and the separation of the body from culture that is the problem or paradox that concerns a number of feminists who have become aligned with the emergence of corporeal feminism. As Braidotti argues, this 'left all issues related to bodies, pleasures, eroticism and the specific ways of knowing the human flesh hanging nowhere' (2002: 30). Further to this, there is the problem of the body being reduced to sexual difference ignoring the complex differences that pass into our being and becoming, such as racial differences (Gatens 1996; Ali 2003; Ahmed 1998, 2000) and transgendered and queer differences (Ahmed 2006; Halberstam 1998, 2005; Butler 1993, 2005). There are a number of concepts that have been introduced to attempt to collapse this dichotomy and rematerialize the body in culture. These include the concepts of performativity and feminine becoming that we will explore in the next section. Neither of these concepts view the body as a fixed 'biological foundation' (Howson 2005: 55) and attempt not to displace 'bodily materiality with social materiality' (ibid.: 71). The orientation of this work relates back to some of the questions and problems with which we started this chapter concerning the relationship between the cultural inscription of difference and embodied subjectivity. As Schiebinger (2000: 3) argues, the 'difference dilemma' is a major part of studies of the body and of body theory. If de Beauvoir (1953) was apposite in her focus on feminine becoming, then the question becomes one of exactly how this process of becoming occurs. As we will see, the concept of becoming has very different meanings and is utilized in rather different ways to understand processes of subject formation. The body, although making a return, appears in rather different forms throughout this work and as such raises further questions about the relationship between materiality and affectivity.

FEMININE BECOMING AND INTERNALIZATION

One interpretation of the concept of feminine becoming as introduced by de Beauvoir can be found in feminist work that draws explicitly on sociological concepts that we investigated in Chapter 1. We briefly reviewed some examples of this work that link the cultural disciplining of the body, in relation to cultural norms, with a process of *internalization*. This process is seen to be lived out and enacted by female subjects in ways that are often detrimental to their own health and well-being

(Bordo 1993). The work of Bordo and others within the tradition explicitly draws on Foucauldian concepts such as the *docile body*, *bio-power* and *micropractices of self and social regulation*. This is a 'cultural approach to the body' (Bordo 1993: 35) that explores how cultural norms become turned on the self creating forms of self-surveillance and self-practice. In extreme and exaggerated forms these can result in practices such as anorexia and bulimia. As Bordo argues, 'denying oneself food becomes the central micro-practice in the education of feminine self-restraint' (1993: 130).

We saw in Chapter 1 how one of the key aims of sociologists of the body was to collapse and even transcend some of the dualisms that have been central to Western thinking on the question of what makes us human. The problem as Turner (1984: 248) suggests is to overthrow a 'number of perennial contrasts' between, for example, structure and agency, mind and body, nature and will and the individual and society and offer solutions that are neither deterministic nor view the body as somehow existing prior to social and cultural processes. Bordo (1993) makes an interesting case for examining how these contrasts are embedded and circulate in social practices in such a way that they are difficult to overthrow and transcend. One of the key contrasts, as we have seen throughout the book, is mind–body dualism, or what is also known as Cartesian dualism. In Bordo's account of this dualist axis, she explores how particular distinctions between the mind and body might account for the higher incidence of eating disorders in relation to feminine becoming. She argues that because the 'thinking self' is associated with the mind (masculinity), the body becomes constituted as '"alien," as the not-self, the not-me' (Bordo 1993: 144). Within this dichotomous way of specifying corporeality she argues that hunger can become constituted as 'an alien invader' (ibid.: 146) with thinness (as a practice of self-regulation) elevated as a 'triumph of the will' (ibid.: 147). Bordo argues that this distinction between the mind and body is one that is part and parcel of 'gender ideology' (ibid.: 110). It circulates endlessly across social practices that position women in relation to their bodies rather than 'intelligence and forethought' (ibid.: 2). As she suggests, this dualism is not just a philosophical argument but a 'practical metaphysics that has been deployed and socially embodied in medicine, law, literary and artistic representations, the popular construction of self, interpersonal relationships, popular culture and adverts' (ibid.: 13).

Bordo's focus is on how to understand and interrogate the relationship between a particular politics of representation of the female body and the kinds of body practices that women are increasingly turning to in liberal democracies to feel better about themselves. This might include compulsive exercising and dieting and cosmetic surgery, which are participated in with the aim of increasing and achieving a sense of success and satisfaction across various aspects of women's lives. This is heightened in a climate where the belief in self-determination has become a normalized aspect of life due to the rise of the psychological sciences and the fiction of the autonomous self (Rose 1989, 1996). The practices that have become normalized as part of feminine becoming, according to Bordo, are those that enable 'control of the unruly body' (1993: 149) and the mastery of will with accompanying psychophysical pain. Thus the body, for Bordo, is the site of the reproduction of particular cultural norms of femininity that become *internalized*, forming the basis of particular practices of self and social regulation. Although Bordo's analysis is very much based upon the politics of female representation which reproduces the mind–body dualism in particular ways, she argues nonetheless that her approach does not simply reduce the body to a textual or discursive effect.

CORPOREAL FEMINISM

Although Bordo argues that the body is 'a practical, direct locus of social control' (1993: 165) the extent to which her approach moves away from the discursive production of the body has been subject to debate (Howson 2005). One problem with the reliance on Foucauldian concepts is the exclusion and lack of attention to a realm of bodily affectivity that is increasingly becoming a central aspect of body theory. As we saw in the opening of this chapter, cultural norms are not simply internalized but engender a background of felt orientations to the world that are not easily verbalized or understood. It is this realm which often discloses embodied experience, and which needs a more complex understanding of self-formation than internalization can invoke and suggest. As we will see, it is for this reason that many feminists have turned to psychoanalytic concepts to theorize 'internalization' and feminine becoming, preferring to move away from sociological concepts and to embrace a philosophical re-reading of psychoanalysis in the context of the problems with cultural inscription (Grosz 1994; Braidotti 2002).

> Philosophy has always considered itself a discipline concerned primarily or exclusively with ideas, concepts, reason, judgement – that is, with terms clearly framed by the concept of mind, terms which marginalize or exclude considerations of the body... As a discipline, philosophy has surreptitiously excluded femininity, and ultimately women, from its practices through its usually implicit coding of femininity with the unreason of the body. (Grosz 1994: 4)

Elizabeth Grosz and Rosi Braidotti are two seminal feminist writers, both philosophers by training, who are part of what has come to be known as 'corporeal feminism' (Howson 2005). We will start the discussion of this term with a consideration of the writings of the Australian feminist Elizabeth Grosz (1994). She argues against the 'somatophobia' (a fear of the body) that is found, she maintains, in philosophy and feminist work that originates within a social constructionist paradigm. In line with work more generally within body theory, Grosz outlines her commitment to transcending dualistic thinking on the body. As we have seen, this is a key theme of work across the humanities, which is attempting to develop a range of body concepts that can do just this. The body needs to be radically reconfigured, she argues, if we are to understand how cultural, social and historical forces work to transform it. This is, as you will be aware, often framed as the problem of cultural inscription or materiality. We have already observed that dualistic thinking tends to work in binaries – mind–body, reason/passion, for example – in which one pole of the binary takes up a negative and inferior status at which actual subjects, such as the working classes, people with different sexualities, colonial subjects and women, are usually positioned. Feminine becoming, for Grosz, is not the expression of 'a pre-existing ability or compound' (1994: 10), but rather denotes the ability of bodies to 'be affected by other bodies' (ibid.: 12). Although this might point towards the very socially inscribed body that is central to social constructionist thought Grosz argues that this is not so. One of the problems for body theory, she holds, is the paucity of languages that allow us to develop 'non-dichotomous understandings of the body' (Grosz 1994: 20). Social constructionism, she contends, actually relies upon the mind as the site of interpellation and transformation (as with the work of Foucault), and therefore assumes the workings of consciousness in processes of subject formation. Grosz turns to psychoanalysis to develop a model of the *psyche* or *psychical* to explore our more felt, lived relations to ourselves and others.

Thus, body image, for Grosz, is not a visual map or cartography of the body (the body as it is looked upon in the mirror, for example), but rather a collection of 'felt intensities' that are derived from bodily sensations. We are required in Western liberal democracies to experience ourselves as coherent, whole and unified – this is captured by the psychoanalytic concept of the *ego*. However, this sense of 'corporeal wholeness' (Grosz 1994: 32) has a 'fantasmatic dimension' (ibid.: 38) as it is based upon a fundamental misrecognition or illusion. We are required to have a sense of self and distinct body boundaries that deny or silence the more porous and permeable aspects of our embodied experience. We explored this at length in Chapter 2. Our body image is thus formed through our encounters within a complex social space that requires us to 'hang together' an image that is cross-cut by different 'logics and rhythms' (Grosz 1994: 105). These different logics and rhythms are those that are felt through what Grosz terms the 'lived body' and which generate intensities that are the traces of the links which are denied and obscured. In other words, there is a complex relationality to the cultural inscription and materialization of the body that fundamentally links the body to other practices, bodies, entities and so forth. So the body, for Grosz, is 'a site for the circulation of energetic intensities' that might be difficult to see and verbalize (1994: 138). We can see here a return to understandings of the body that are not based upon separation but connection. The question of where your body ends and the other begins is now much less certain and clear cut. The following sums up the orientation of Grosz's contribution to corporeal feminism and is worth quoting at length: 'the body as a discontinuous, non-totalisable series of processes, flows, energies, speeds and durations, may be of great value to feminisms attempt to re-conceive bodies outside the binary oppositions imposed on the body by the mind/body, nature/culture, subject/object and interior/exterior oppositions' (1994: 164).

Like her contemporary, Rosi Braidotti (2002), Grosz experiments with concepts from Deleuzian philosophy to 'think' the body as a non- dualistic process rather than a substance. We will spend some time in Chapter 5 exploring Deleuzian concepts. In this chapter we will explore the general assumptions of some of this work and particularly its commitment to 'thinking the body' outside of rigid demarcations and boundaries. This relates to the particular way in which the concept of becoming has been adopted and utilized, and so we will spend some time exploring this further. We have seen in this chapter how social difference tends to be articulated through distinctions between the normal and the other. In other words, although bodies are inscribed through a complex relational matrix, this matrix is cross-cut by power and ideological processes that positions bodies in particular ways. What this focus on normalization misses, according to Braidotti (2002), are the contradictory, contested and multilayered ways in which bodies are inscribed. Like Grosz, Braidotti introduces a notion of bodily affectivity to point to a realm of felt experience existing at the intersection of social normativity and the body's capacity to 'fight back'. Thus, Braidotti, like Grosz, argues that what are important are *affect*, *desire* and *imagination*, and how they are organized, channelled and transformed. The question for Braidotti is how we string together a sense of self, 'under the fictional unity of an I' (2002: 22), in the face of 'power, struggles and contradictions' (ibid.: 25). This question is one that has been central to the work of another feminist philosopher who has also addressed the question of the place of the body in the cultural inscription of difference: Judith Butler. Butler is also a philosopher by training and has also turned to psychoanalysis to theorize the becoming subject. However, Butler is more interested in normalization: how subjects are produced as sexed subjects in such a way that they experience gendered distinctions as natural, normal and inevitable.

PERFORMATIVITY

Butler (1990, 1993) has developed the concept of gender *performativity* to capture the process through which bodies are materialized as sexed bodies. She brings together a Foucauldian approach to the discursive production of the body with psychoanalytic concepts. As Braidotti (2002: 40) argues, the supplementation of Foucauldian concepts with psychoanalysis provides a way of addressing the 'discursive glue' that holds the subject together. Subjects do not simply internalize the positions that they are invited to inhabit (for example, the cultural norms of masculinity and femininity), but, rather, struggle with the contradictions. These struggles occur at a non-conscious or unconscious level, according to Butler, and are managed through the deployment of certain defensive strategies such as projection, or even through experiences of melancholia (a sense of loss and lack). The contradictions also have the potential to create gaps, silences and disavowed identifications that can be inhabited and enacted. Thus the gaps might produce resistance to gendered norms, and even the formation of what Foucault termed 'reverse discourses' through which people experience their embodied subjectivities. One of the key focuses of Butler's work is exploring examples of subjects who refuse gendered norms either through parodic irony or mimicry (drag), or through playing or mixing the binary of masculinity and femininity to the extent that it is difficult and sometimes impossible to make such a distinction. In her book *Bodies that Matter* (1993) she focuses upon a documentary directed by Jennie Livingstone that documents the 'voguing scene' amongst poor black and gay Puerto Rican men living in New York. The vogue series of dance movements was popularized by Madonna in the 1990s and is based upon exaggerated movements taken from the catwalk. Livingstone's work *Paris is Burning* (1990) is a visual display of the mimicry of these movements in the context of the 'Ball'. The Ball takes place once a year and provides a space where these men can live out imaginary scenarios in which they are superstars, celebrities, adored and adorned and applauded and judged by their peers for their costume, comportment and embodiment of such moves. This stands in sharp contrast to the very real realities of their lives, which, as we discover in the documentary, are governed by pain, hardship, adversity, oppression and often violence, and even the avoidance of death on the streets.

This reminds us how transformations or resistance to cultural norms are not simply due to acts of will or volition. They are governed by complex unconscious factors and social fantasies that are related to a subject's own personal histories and how these intersect to produce 'the bundle of contradictions that is the subject' (Braidotti 2002: 39). Braidotti argues that although cultural norms act like magnets 'drawing the self in certain directions' (2002: 40), we do not simply internalize them in any straightforward fashion. Any changes or transformations might incur pain and a complex 'working through' that, as we have seen throughout this chapter, are registered and disclosed through forms of bodily affectivity that point towards a *felt* orientation to the world. Braidotti's use of the concept of becoming via Deleuze is therefore both similar and yet strategically different to Butler's concept of performativity. She laments Butler's focus upon loss and melancholia as the cost of taking up normalized subject positions (Braidotti 2002). However, although she is keen to recognize the power of normalization in shaping bodies, she does not believe that loss and melancholia describe the pain of transformation. We exist, Braidotti argues, in a plenitude of possible becomings that are continually changing and transforming. The intensities that this engenders create pleasure and affirmative and joyful affects that open the subject up to a 'multiplicity of possible differences' (2002: 71). She holds (Braidotti 2002: 77) that there are a 'multiplicity of sexed subject positions' that allow for different possibilities of becoming female or becoming woman that cannot be contained by the binary logic of masculinity and femininity. As she argues, the 'nomadic or intensive horizon is a subjectivity 'beyond gender' in the sense of being dispersed, not binary, multiple, not dualistic, interconnected, not dialectical and in a constant flux, not fixed (Braidotti 2002: 80). However, the capacity to engage in alternative becomings is never due to volition or will, judgement or choice and, therefore, for Braidotti, opens up investigations of the body to a complex realm of affectivity that is little known or understood, but is *felt* in a very real and profound way. The differences and similarities between these feminists and the debates that have ensued are far from resolved, but all point towards the body as a process, rather than a substance, and to the importance of examining subjectivity. The concept of subjectivity draws our attention to the complex processes through which subjects construct a liveable sense of self in the face of multiplicity, ambivalence, contradiction and inequalities and oppressions (see Butler 2004, 2005).

Conclusion

> Western culture privileges talk and text as legitimate forms of communication, yet there may be no words or language to name and communicate certain forms of experience, particularly physical sensation and its contribution to subjectivity.
>
> Howson, *Embodying Gender*

This chapter has covered various approaches to the cultural inscription of bodily difference that all converge, in different ways, around the question of bodily affectivity and its importance for understanding the body and embodiment. In different ways they all cite the importance of non-linguistic forms of communication and how the body is never a singular body but a complex relational process. Our experiences of the world, according to the perspectives that we have covered, are 'known through the 'bodily mode' (Howson 2005: 148). This is not a separation of the body from the social, but rather a reconfiguration of the body in which the dualisms of individual and social, mind and body, nature and culture and even the inside and outside are collapsed and viewed as complex relational processes. What is specific across Western liberal democracies in the early twenty-first century is a belief in self-determination and a cultural injunction to live our bodies as singular, bounded and clearly separated from others. It is the paradox and contradictions that this creates that form the backdrop to many of the studies covered in this chapter. In Chapter 4, 'Lived Bodies', we will turn to studies that take the body and embodiment as their particular focus and specifically explore those across the humanities that have examined the more 'physical' questions of the body. This will include an examination of 'body parts' such as the mouth and teeth and the body in health and illness. The chapter will further illustrate how a focus upon the *lived body* troubles the idea that the biological and the cultural are two separate, two absolutely discrete entities that somehow interact.

4 LIVED BODIES

INTRODUCTION

The concept of the 'lived body' brings together a variety of different perspectives within body theory that start with our lived, subjective experience of corporeality. This might include our lived experience of the body as it becomes known to us in health and illness, through our sensual experiences or in relation to 'body parts' such as the mouth and teeth, for example. The concept of the lived body unites perspectives that go beyond exploring how bodies are represented to instead ask and interrogate how we 'live' our bodies. The perspectives are distinct, drawing from some of the different places and disciplinary perspectives that we have already encountered in the book so far. They are united by their commitment to explore the 'lived body' as neither having prior historical and cultural existence nor being reducible to some kind of fixed essence (that is, human nature). Theorizing takes place 'from lived bodies' (Williams and Bendelow 1998: 3) and assumes that bodies are always 'unfinished' and in process. The focus is on *experience* and how we might account for the specificities of our material existence without presuming that materiality can be easily separated from social and cultural processes. There is also a key commitment across this area of body theory to refuse thinking the body in binary terms, such as the separation of the mind from the body, for example. Unlike naturalistic accounts of the body that we encountered in Chapter 1, the bodily basis of experience is not considered to be un-changing and requires what Williams and Bendelow frame as a more sensitive engagement with the 'problem of biology' (1998: 17). The 'problem of biology' refers to the commitment of scholars across the humanities to move beyond a 'social constructionist' paradigm (see Chapter 1), and reconsider the materiality of bodies in new and exciting ways.

One definition of the lived body that we considered in Chapter 3 was that which focuses upon the 'kinesthetic lived-bodily incorporation of the sense of the world' (Charlesworth 2000: 64). The concept of kinesthesia refers to a body which is sentient and which moves and engages with the world through a form of *corporeal consciousness*. In other words, perception (of the world) is not cognitive, whereby thinking is separated from the body and located within the mind, but rather occurs through a 'thinking' body, which is seen to have particular kinds of intelligences and competences. This is one way in which the lived body has been brought into body theory. One aspect of this work is a reconsideration of the importance of the senses for understanding bodies. Within these reformulations, as with much work across body studies, bodies are viewed as having the capacity for thinking and experiencing in ways that challenge Cartesian dualism – that is, a separation of mind from body. This work will be developed in the first section of this chapter by considering some of the new concepts that have been introduced to understand the senses and sentient body for framing relationships between the body and social and cultural processes. One of the key focuses of this work is on *movement* rather than viewing the senses as fixed, interior processes marked by their location or place within the body.

THE SENTIENT BODY

It is often taken for granted within Western cultures that our sense of bodily awareness is primarily structured through five senses: touch, taste, smell, hearing and vision. Although there is much discussion about whether there is a range of peripheral senses, such as proprioception, which govern our movement through the world, it is generally agreed that the world impinges upon us and is actively perceived through a combination of this primary sense organization. Although the senses are often discussed as separate processes it is now agreed that they work in combination and communication with each other rather than as isolated forms of bodily awareness. The term that is used to describe these networked connections and processes is synesthesia. This is what Sobchack characterizes as a form of 'cross-modal sensorial exchange' (2004: 69). However, discussion of the senses as enabling and providing the body's capacity for awareness and movement has been hampered by a number of assumptions that have taken on the status of historical truth. The first that is important for our consideration of the senses is one which is related to Cartesian dualism. As we have seen throughout the book so far, Cartesian dualism assumes that the mind and body are separate entities, with the mind generally being seen to be the site of thinking and reason and the body produced as a machine-like substance with its own physiological processes. This distinction, which views the mind as having a set of intelligences separate from the body, is mirrored in some of the assumptions made about the senses. Pasi Falk (1994) has argued that in industrialized cultures the senses are understood and made intelligible through a hierarchy from the higher to the lower that mirrors the mind–body dualism. We will consider Falk's discussion in more depth as it is important for understanding some of the reformulations of the senses that have challenged this hierarchical distinction.

Falk argues that vision and hearing (aurality) are considered to be the higher or 'distant' senses, most closely aligned with reason, thought and reflection. The term 'distant' is used to denote their supposed distance from the body which is seen to root the human subject in more primitive forms of bodily awareness. Vision and hearing are those senses that enable the subject to dominate and transcend their supposed animality and transform the world accordingly. The lower or 'contact' senses are those that are seen to be located in the body (considered to be separate from the mind) and that provide forms of bodily awareness which are considered most distant from processes of rational reflection. The 'lower' or contact senses include taste, touch and smell and are considered more brute, direct and vulgar providing our supposed link to our animal heritage. Within a naturalistic body paradigm (see Chapter 1) these process are viewed as lower in terms of our evolutionary development, and as those that ideally we are less reliant upon in our move to more civilized modes of conduct, experience and behaviour. The rational, civilized subject is judged to be a subject who undertakes considered reflection rather than being 'swept away' by smell, taste or touch. This mirrors mind–body dualism, which assumes that the mind is the site or citadel of reason with the body produced as a brute, inert substance that is ideally brought under the control and domination of the mind. This dichotomy is nicely summed up by Maxine Sheets-Johnston (1999) who argues that we assume that it is the mind that 'thinks' and the body that 'does' (as a result of thinking). But what would happen if we were to refuse this sense organization and the attending mind–body dualism upon which it relies? Contemporary work across body theory is doing just this, reintroducing and reformulating bodies as having the capacity to both affect and be affected, with the result that the mind and body are not considered in binary terms (Williams and Bendelow 1998).

TOUCH

We tend to think of touch as one of the most direct, physical senses that binds us and connects us to others. From the caregiver's touch to their infant, or the touch expressed by lovers through the caress and hug, touch is a form of tactile communication that shows our primary interconnectedness with others. However, although tactile communication is often framed as a physical sense, there are other versions of tactility that exist in our lexicon of touch that point towards its production as a rather different kind of sense. Let us consider the senses of touch that we might experience by 'moving in time' with others. In Chapter 1 we reflected on the concept of muscular bonding, which refers to the kinds of affective or emotional experiences that are often produced when people move together rhythmically in time (McNeill 1995). This might be through coordinated activities such as dance, various drills (such as in marching), or through the structured group expression of practices such as Tai Chi or Qigong. The felt visceral sense that is often achieved and experienced through moving together in synchrony is one that is aligned to a sense of well-being and expansiveness. It literally feels good and points towards understandings of being touched and touching others that cannot be captured by a reduction of this sense to a brute, literal, physical sensation. As Finnegan argues, 'the experience of working, marching, playing, loving or competing together, "in sync" is a real one in human interconnectedness, even without actual "touch" in the literal sense' (2005: 22). In a similar vein we might talk about being 'touched' by a film, which Sobchack describes as the 'carnal sensuality of the film experience' (2004: 56). These experiences of touching and being touched point towards a sensual feeling that cannot be understood in a literal sense. This is a sense of bodily awareness that is not simply a physiological reflex action rooted in an inert, brute body.

David Howes (2005) develops this notion of touch as a different kind of bodily knowing, obscured or silenced by Cartesian dualism, through the concept of *skin knowledge*. 'Skin knowledge' refers to a form of intelligent bodily knowing or under-standing that forms an important component to our sense-making activity. Rather than touch being viewed as a contact, lower sense, it is understood as a form of sentient activity that provides body competences that enable our movement through the world. The skin is not simply a physical protective covering but one that also creates the possibility of a different kind of bodily knowing. It exists as an interface between the self and other, biological and social and organic and inorganic, and is both internal and external. It acts as a bridge or 'intermediary screen' (Anzieu 1989: 4) between the psyche and the body making it the primary site or instrument of interaction between the self and other. The skin is therefore an instrument of communication that allows us to sense and feel in the world. Didier Anzieu was a psychoanalyst who worked on a dermatology ward and developed the concept of the 'skin ego'. He argues that the skin is a sense organ that, 'is the most vital: one can live without sight, hearing, taste or smell, but it is impossible to survive if the greater part of one's skin is not intact' (1989: 4). Anzieu's concept of the 'skin ego' has much in common with the concept of skin knowledge.

Howes illustrates his concept of skin knowledge by making visible under-standings of the skin as 'knowing' and intelligent that have existed historically, although perhaps becoming marginal or alien to mainstream industrialized forms of 'knowing'. He discusses the nineteenth-century naturalist Henry David Thoreau and his 'seemingly unconscious ability to find his way home to his cabin in the woods in the dead of night' (2005: 27). In this example the body is seen to possess a different form of intelligent thinking that is felt and sensed rather than verbalized and articulated through language or cognition. The following passage from Thoreau is reproduced by Howes and crystallizes the importance of skin knowledge in our interactions with the world and others.

> It is darker in the woods, even in common nights, than most suppose. I frequently had to ... feel with my feet the faint track which I had worn, or steer by the known relation of particular trees which I felt with my hands... Sometimes, after coming home thus late in a dark and muggy night, when my feet felt the path which my eyes could not see, dreaming and absent-minded all the way, until I was aroused by having to raise my hand to lift the latch, I have not been able to recall a single step of the walk, and I have thought that perhaps my body would find its way home if its master should forsake it. (Thoreau 1968 in Howes 2005: 27)

Howes (2005) suggests that modern, urban industrialized living has obscured these forms of 'knowing' that have increasingly been replaced or covered over by more mechanized ways of knowing. These might include the watch or clock, automated signal systems, and screens of all kinds which mediate and acculturate our interactions between the self and other (see Harbord 2007 for a discussion of screen cultures). He terms these new ways of 'knowing' 'electronic skins' that produce the body as a machine connected up with other automated and digital technologies so that the sensing body is placed in the background. These electronic skins create and shape different ways of 'feeling in the world' (Classen 2005: 402). These processes of 'feeling in the world' are both disembodied, in the sense that they rely or forge physical detachment, whilst simultaneously overloading and overwhelming us with visceral stimulation. The kinds of skin knowledge that Howes identifies are not simply cognitive forms of reflexivity, but tactile forms of knowing that *attune* us so that we are permeable and open to being affected by the other, human and non-human. This takes us back some way to work that we were exploring in Chapter 2 which examined theories which suggest that permeability and connectedness rather than separation and self-contained individualism are what defines our encounters with others.

This re-evaluation of touch as a form of *intelligent knowing* shows us how the sentient lived body is a good starting point for examining the complex intersection of nature with culture, the individual with the social and the psyche with the somatic. The concept of skin knowledge and Anzieu's concept of the skin ego both assume that the skin provides a site where there is a continual exchange and interchange between what we might understand to be the inside and the outside. The very fact that the skin is simultaneously internal and external, as well as permeable and impermeable, points towards the dangers of thinking the senses through binary terms such as the mind and body. What this discussion of touch, through the concept of skin knowledge has disclosed, is the very strong interdependent relationships between the psychic, corporeal and the cognitive to the extent that they should be considered thoroughly entangled processes.

TASTE

Taste is another sense that is considered more brute and 'physical' in its manifestations, although it is also recognized that we have different habits and practices in relation to food and its consumption. Falk (1994) takes both these ways of framing 'taste' in his study of orality and its cultural and historical organization. We tend to think of orality as the study of talking, of oral cultures, but orality also refers to our sense of taste and discloses, as we will see, the very close alliance between talking and tasting. Within a naturalistic body paradigm taste might be equated to the action of taste buds on the tongue which allow us to discern certain qualities of food: sourness, saltiness, bitterness and sweetness, for example. However, we also have a number of terms that show how the tongue is also closely aligned to speech. The idea of being 'tongue-tied' shows how the tongue is involved in communication and how the mouth, as the 'body part' which contains the tongue, is itself an over-determined body part or orifice. We also use the term 'mother tongue' to refer to languages and vocabularies that might have been lost or silenced by the global dominance of certain languages, such as English. This might have occurred as part of the rise of imperialism and colonialism, for example. We might also talk about 'whetting the taste buds' as a way of signalling your newly acquired taste for certain activities that might not be considered 'biological' in any direct sense. This might include a newly acquired taste for an activity that you have sampled and decided you want to continue with in a more committed way.

We also talk about 'educating' the taste buds which points towards the role of *training* and *discipline* in relation to our food habits and tastes. Indeed, industries such as the perfume industry and the wine tasting industry are built on the premise that people can be educated to acquire a sense for discerning subtle and nuanced flavours or smells that do not correspond to sourness, saltiness, bitterness and sweetness, for instance. This sense of taste goes beyond its constitution as a 'contact' sense and instead views the acquisition of certain tastes as expressions and manifestations of social distinctions, such as bourgeois middle-class culture. This was a key focus of the work of the French sociologist, Pierre Bourdieu (see Chapter 3) and is captured by the statement; that person has 'no taste' or 'bad taste' with the attendant connotation that they are perhaps lower, more inferior and even primitive. Something or somebody might also leave you with a 'bad taste' in your mouth, which is a way of referring to the passing of feeling, emotion or affect that has been left with you and which it is felt does not belong to you. The mixing of the biological with the cultural and psychological is captured by the possible cure for such an experience, which might be a drink to 'wash' it away.

THE MOUTH

This brief consideration of some of the lexicon of terms which surround 'taste' illustrates how our sense of taste cannot simply be understood as a brute, physical sense, ultimately grounded within particular biological understandings of the body (the taste buds, for example). With this in mind the anthropologist and cultural theorist Pasi Falk has taken 'taste' as an object of historical and cultural inquiry. His starting point is the lived body and its experiences of taste in very different historical and cultural contexts. Falk's study is considered a good example of approaches within body theory that focus upon the 'historicity of the body' and the effects broader sociocultural shifts and changes have on 'the *experiential* and expressive aspects of the body' (Williams and Bendelow 1998: 48). Falk's discussion of taste and food consumption is linked through an examination of the 'mouth'. In our earlier discussion of the close link between taste and talking we began to see how over-determined the mouth is as a body part. The mouth connects and links a number of disparate functions, which include eating, drinking, biting, talking, sucking, kissing, smiling and shouting, for example. These functions define what a mouth is capable of doing and show how as a body part it connects up and intersects with sexuality as well as communication and eating. It is an orifice or 'body opening' that is neither entirely inside or outside. In this sense it shares qualities with the skin, in that it is argued that the mouth is an enfolding of the inside with the outside, operating on the boundary of what we might take the inside and the outside to be. It is also an entry and exit point, capable of ingesting and expelling, disturbing again the boundaries of what we might take the inside and the outside to be. We can see, then, that the mouth raises broader questions about the body and how to understand the function and role of specific body parts within broader social and cultural processes.

Falk (1994) distinguishes two very different 'mouths' which he suggests are emblematic of broader sociocultural relationships. The first is the 'collective mouth', which he ties to the existence of pre-modern cultures in which distinctions between the self and other and the 'me' and the 'not-me' are drawn at the level of the collectivity rather than the individual. In other words, what is considered and constituted as 'Other' might be the supernatural or nature, rather than another person. In this sense there exists a more collective or group sense of subjectivity, which is mirrored in the codes and regulations that govern food and its consumption.

Falk contrasts the symbolic function of the ritual meal with the function of the 'shared meal' within advanced liberal cultures. The ritual meal is a shared meal which operates as a 'collective constituting ritual' (1994: 25). There are strict codes and taboos which govern what can be eaten, by whom, when and how. Everybody literally knows their place within the community, and this is reflected in who sits where, who is allowed to eat what, what is considered good, bad and inedible and in what manner the food is consumed. The key principle that governs the function of the ritual meal, according to Falk, is that there is little or no room for *matters of individual taste*. This is distinctly different to the function of the 'shared meal' within advanced liberal cultures. We might see traces within modern cultures of the ritual meal, which is often associated with religious ceremonies. These might include shared meals during the Jewish Passover, Christmas or Thanksgiving, or the ritual meal that follows fasting during the Muslim month of Ramadan. However, Falk suggests that the more ritualistic aspects of the shared meal have been subsumed by matters of individual taste. He terms this a shift away from the 'collective mouth' to the 'individual mouth'.

The 'individual mouth' constitutes 'a shift to modes of social interaction in which individuals acknowledge – at least in principle – each other's autonomy as exchanging and/or communicating subjects' (Falk 1994: 29). The link between people within the context of the shared meal is what is said, (talk), rather than what is eaten. This is an orality that Falk suggests presupposes separateness and individuality. We have already seen how, within our lexicon of terms surrounding orality, talk and taste are closely aligned. Because of this close alignment, Falk suggests that the French surrealist film-maker, Luis Buñuel, in his comedy, *Le Fantôme de la liberté* is able to invert the oral and anal functions of the digestive process so that interaction and speaking become totally independent from the matter of ingesting. In a scene in the film, guests arrive at the host's house apparently to participate in a shared meal. The table is set in such a way that we might expect a dinner party to ensue with each person taking their place. However, we find that each person sits upon a toilet, first making sure that the lid is up, provided with their own toilet paper and reading materials if needed. They then begin to talk and communicate with speech operating as the primary function binding them together. When they need to eat they ask in hushed and discreet tones where they might go, and find their way to private, individual quarters where they eat in silence. Falk suggests that Buñuel uses this as a surreal device to make visible how eating has become an individualized affair. Falk links this shift from the collective to the individual mouth to a broader sociocultural shift from a 'collective self' to a self-contained, atomized sense of individuality. We have explored this in different ways throughout the book so far, but it is worth quoting at length from the British sociologist, Nikolas Rose, who has written extensively on the emergence of this new culture of the self, which he refers to through the concept of the *fiction of autonomous selfhood*.

> I want to suggest that the *relation* to ourselves which we can have today has been profoundly shaped by the rise of the psy disciplines, their languages, types of explanation and judgement, their techniques and their expertise. The beliefs, norms and techniques which have come into existence under the sign of psy over the last century about intelligence, personality, emotions, wishes, groups relations, psychiatric distress and so forth are neither illumination nor mystification: they have profoundly shaped the kinds of persons we are able to be – the ways we think of ourselves, the ways we act upon ourselves, the kinds of persons we are presumed to be in our consuming, producing, loving, praying, sickening and dying. We need to abandon the belief that we are 'in our very nature' discrete, bounded, self-identical creatures, inhabited and animated by an inner world whose laws and processes psychology has begun to reveal to us. On the contrary, we are 'assembled' selves, in which all the 'private' effects of psychological interiority are constituted by our linkage into 'public' languages, practices, techniques and artefacts. (Rose 1996: 250)

THE MOUTH AND TASTE

Rose suggests, like Falk, that our sense of ourselves as separate and self-contained exists because of the ways in which we are linked and connected through our bodies to practices which address us 'as if' we were selves of a particular kind. What we see in relation to eating is that taste becomes increasingly a matter of individual choice and preference, rather than being constituted by collective codes and ritual taboos. However, this does not mean that taste is no longer regulated. Although we might experience our expression of taste as an individual matter, Falk shows how tastes are disciplined, educated, shaped and trained through moral and scientific discourses which govern what is good and bad for you. The emergence of a whole consumer industry surrounding nutrition and dietetics illustrates the salience of new codes that govern taste in relation to the 'individual mouth'. These discourses have helped to inaugurate distinctions in taste and food habits which constitute certain habits as individual pathologies which need to be corrected. These include over-eating, gluttony and obesity as well as eating disorders based on self-starvation or binging and purging. What receives less attention in the constitution of certain habits as *individual pathologies* is the role that the advertizing and marketing industries play in creating and reproducing contradictions that individuals are required to manage through their own eating practices. These include the marketing of so-called naughty foods, such as ice-cream and chocolate, which are made to appeal to a sense of guilt or naughty pleasure. This is constituted as a kind of sensual or bodily pleasure which is separated from our 'rational' awareness that these foods may not be good for us or have little nutritional value. In other words, there is an inherent mind–body dualism marketed back to us in the advertizing of snack foods which aim to appeal to an economy of emotions such as guilt, shame and even weakness (of will-power). The body that is represented to us is one, which, through taste, might override rationality and provide us with a more direct sensual awareness. We are back with the salience of taste as a brute, direct and sensual bodily awareness.

So far in the discussion of the two 'contact' senses, touch and taste, we have seen how humanities scholars have reformulated the senses as forms of bodily awareness that cannot be separated from the broader operation and workings of social, historical and cultural processes. Although the 'lived body' is the focus of this emerging trend within body theory, there is always the danger that such work retreats into the very social constructionist paradigm they are at pains to avoid. As Williams and Bendelow (1998) consider in their review of work that takes the 'lived body' as its subject, there is always the danger of ending up with a very over-socialized conception of the body in which the materiality of the body is elided. More focus is given in Falk's work to the effects of broader sociocultural shifts on how we experience and express our food habits and tastes, which he links to the increasing individualization of the self through the emergence of a new way of specifying the mouth. In the next section we will consider work on taste and the mouth which deals more expressly with the more visceral aspects of taste. This work considers the concept of *abjection*, which refers to the deep sense of revulsion, disgust and horror we might have to particular foods and particular actions which accompany what a mouth is capable of doing. The experience of abjection is literally felt in and through a bodily awareness that might include feelings of nausea or vomiting or the feeling of being physically repulsed

---

Case Study

Imagine spitting into a glass and then drinking it. How might that make you feel? How might it make you feel to watch somebody else perform this act? If you cannot imagine it, try doing it yourself or asking a friend to do it. Now imagine drinking your friend's saliva! This is exactly what Gordon Allport (1955), the American psychologist did in a series of experiments

that he conducted to explore what we would now term the experience of *abjection*. Allport was interested in how our own bodily fluids, such as spit and saliva, could, in an instant, upon being expelled, become 'not-me'. Once outside the boundary of the body our own bodily fluids, he suggests, become a site for the experience of disgust and revulsion. As he argues, 'That which is spit which might prompt us to attempt to move away from something or someone. In the next section we will consider the role of *abjection* in relation to our sensual bodily awareness by examining the fragility and precariousness of what we might take to be inside and outside in relation to the constitution of the boundaries of the body.

ABJECTION

> Think first of swallowing the saliva in your mouth, or do so. Then imagine expectorating it into a tumbler and drinking it! What seemed natural and 'mine' suddenly becomes disgusting and alien.
>
> Allport *Becoming: Basic Considerations for a Psychology of Personality*

In the book *Powers of Horror* the French feminist psychoanalyst Julia Kristeva (1982) describes the abject as that which 'disturbs identity, system and order ... [and] does not respect borders, positions, rules'. It is 'that which defines what is fully human from what is not' (ibid.). Kristeva is referring to that which becomes threatening because it disturbs the bodily boundaries we try to create and maintain between the self and other, for example. The abject is that which is commonly associated with bodily fluids and waste products that leave via open wounds or bodily openings such as the mouth, vagina or anus. This includes excrement, urine, vomit, blood, saliva and pus, for example. The inside is literally turned outside, threatening the very borders and boundaries between the inside and outside that are central to the maintenance of the human subject as a unified, self-contained individual. 'The abject, for Kristeva, is "dirty", "filthy", "contaminating", "waste": a liminal category, which is neither "self" nor "other", "inside" nor "outside". Transgressive in nature, it respects no borders, rules or positions' (Williams and Bendelow 1998: 124).

---

out can never remain the same again' – undergoes a magical transformation so that it becomes in an instant – 'not-me'. 'Once outside, out for good'! (Allport 1955: 43). Once outside, saliva must remain 'not-me'. To trouble this boundary between the inside and the outside, creates the potential for visceral experiences of disgust that point towards the importance of the maintenance of bodily boundaries in relation to our lived experience of bodies. It is these very visceral experiences of disgust and revulsion that are captured by the concept of *abjection*.

The abject is also that which is seen to connect us to what is viewed as more bodily, more animal-like and therefore primitive; what is considered lower, vulgar, defiled and *disgusting*. Disgust has a particular and special relationship to bodily entry and exist points such as the mouth. It plays a key role in identity formation and demarcates a number of key boundaries, such as the inside and outside, the natural and the cultural and the mind and body. In a consideration of disgust as the primary embodied experience of abjection, William Ian Miller (1997) explores how disgust tends to be experienced through feelings of contamination, pollution and/or danger. In relation to the mouth and eating, Miller makes the following observations:

> Once food goes into the mouth it is magically transformed into the disgusting. Chewed food has the capacity to be even more disgusting than faeces. The person who routinely checks the production of his bowels does not have the same type of interest in looking at well-chewed food he has spit out of his mouth: there is no sense that masticated food can be looked on with the pride of creation that faeces can. The sight of chewed food, either in the mouth or ejected from it, is revolting in the extreme. Some parents who have no trouble changing their children's diapers still have to steel themselves before touching masticated food. Even those few foods which by rule are allowed to be withdrawn from the mouth after entering it or acceptably licked by the tongue are dealt with charily. Even lovers must overcome some small resistance to lick an ice cream cone that the other has licked. (Miller 1997: 96)

In this consideration of disgust we can clearly see how the body is seen to connect us with that which is potentially vulgar, primitive, animal-like and seen to be 'biological' rather than cultural in origin. Although we might seek to 'think' the body in non-dualistic terms, we can see in this discussion of the mouth and abjection, just how entrenched the dualism between mind and body is in our experiences of senses such as taste. We can also see how such experiences of abjection are intimately linked to the emergence in advanced liberal cultures of an 'individualized body' (Laporte 2000: x). This body, as we have seen, is engendered or brought into being through two key concepts: individuation and separation. Although body scholars are committed to theorizing bodies as 'unfinished' and always in a process of *becoming*, we need to be equally attentive to the ways in which bodies are made intelligible and therefore potentially lived and enacted across cultural sites and practices. Rosi Braidotti (2002) uses the term *body-culture* to refer usefully to the normalizing ways in which bodies are brought into being across material and cultural practices. Although work across body theory suggests that bodies are more permeable than essentialist approaches to the body have formulated, there still remains a strong commitment, as we saw in Chapter 3, to explore the lived experience of 'the individualization of selves and bodies' (Williams and Bendelow 1998: 89).

SMELL

In this last section on the sentient body we will consider the third sense that is given the status of a contact sense within advanced liberal cultures: smell. In a fascinating book titled the *History of Shit* Laporte (2000) ties the hierarchy of the senses to a hierarchy of waste, showing how, with the rise of the 'individualized self', a fundamental ambivalence was created towards waste products such as excrement. Waste was to become increasingly individualized and confined and dealt with within the confines of the private, domestic space. As Laporte convincingly shows, shit was to become domesticated where, with the linking of sight to reason and rationality, smell was to become disqualified as an object of disgust. Where the other senses are linked to definable body parts or organs, such as the eyes with sight, the ears with vision, the mouth with taste and the skin with touch, smell is much more diffuse. Although aligned with the nose, there is a sense that smell is more porous and fluid, less likely to respect borders and boundaries. As Miller argues, 'smells are pervasive and invisible, capable of threatening like poison; smells are the very vehicles of contagion. Odors are thus especially contaminating and much more dangerous than localized substances one may or may not put in the mouth' (1997: 66).

This might go some way to explaining why excrement and its odour was considered both a site of contamination, but also has increasingly become subjected to what Laporte (2000) terms deodorization or a disinfection process. Laporte suggests that because excrement always carries the '"noxious" trace of the body it departs' (2000: 37), it is imperative that its smell is substituted with the smell of another. This might be the use of essential aromatic oils or fresheners as we see increasingly in the contemporary advertizing for toilet products, for example. Laporte argues that the mythology of contemporary advertizing for toilet products 'contains a compulsive need to eradicate human smell and the "olfactory animal" that man had once been' (2000: 83). As he argues, smell is truly considered to be the bottom of the heap when it comes to the senses, leading to what he terms, 'a disqualification of the olfactory' (2000: 97). Miller makes a similar set of observations in relation to the status of smell and the olfactory sense:

> Smell ranks low in the hierarchy of the senses. That there are bad sights and bad sounds does nothing to undermine the glory of the 'higher senses' of vision and hearing; and that there are delightful fragrances does nothing to raise smell from the ditch. So low is smell that the best smell is not a good smell but no smell at all. And this sentiment predates the twentieth-century American obsession with not smelling. (Miller 1997: 75)

What we can see here is an opposition between the high and the low. Falk (1994) argues that this opposition is thoroughly engrained in the ways in which we account for the status of the senses in relation to our bodily awareness. Smell is associated with the primitive, with bestiality and with the lowly and the inferior.

In the film adaptation of Patrick Suskind's bestselling novel *Perfume* the lead character, Jean-Baptiste Grenouille, has a highly developed olfactory system. His highly attuned sense of smell is aligned in the film adaptation, *Perfume: The Story of a Murderer* (2006), to his birth in the odorous environment of the Paris slums in the eighteenth century. His birth is surrounded by the stench of rotting food and human and animal waste and excrement. Jean-Baptiste is marked as somebody who is 'Other' to middle-class rationality, defined through his bodily sense of smell which is aligned to his lowly working-class beginnings. He is no more than an animal, and yet his way of moving through the world reveals his superior olfactory sense which elevates him above even the most skilled perfumers in Paris. His sense of smell is not only considered primitive but almost other-worldly; throughout the film he is constituted as non-human, filthy and evil. We tend to think of the eyes as the window to the soul, in this case the nose provides the instrument to a world of obsession and compulsion. It is smell and its linking to an animal-like compulsion that ultimately leads Jean-Baptiste to kill young, attractive females in order to bottle and create the ultimate scent that characterizes 'beauty'. In this film adaptation, as in the book, smell is aligned to the primitive and evil, with those succumbing to its odorous compulsion no more than animals. Smell is to the body as vision is to rationality. However, the French sociologist Bruno Latour (2004) perhaps gives us a rather different way of thinking about smell in relation to the sentient body.

THE ARTICULATED BODY

As with many other contemporary body scholars, Bruno Latour is a sociologist who rejects the idea that bodies can be defined as substances or essences. In an interesting article 'How to Talk About the Body?' he defines the body as 'an interface that becomes more and more describable as it learns to be affected by more and more elements' (2004: 206). But what does it mean to describe the body through the concept of 'learning to be affected'? Latour gives the example of what it means to become an apprentice within the perfume industry in France: what does it mean to develop a good nose or smell for odours? We might think of this process as a cognitive process. Indeed the notion of *training* often incorporates the idea that learning is a disembodied practice of knowledge acquisition. That is one of learning to discriminate smells through a process of education, which requires cognitive reflexivity. However, Latour develops a more embodied approach to learning that does not separate the mind from the body, nor assume that learning simply requires a mind housed in a singular body. He argues that what it means to learn is a process that requires the capacity to *learn to be affected* through the conjoining of the body with artefacts, techniques and technologies which define the particular social practice. In the case of acquiring or developing a 'good nose', Latour foregrounds the importance of the body's relationship to the 'odour kit' – a set containing different smells and perfumes. It is the odour kit which *articulates* the body so that it can become more sensitive to the finer and finer distinctions and discriminations which characterize smell. These contrasts, through the conjoining of the body with the odour kit, and the potential of the subject to become sensitive to its own latent possibilities, affect the body with the result that the contrasts become meaningful. The odours now elicit different kinds of potential action and feeling: they have enabled the subject to be transformed. The subject needs the odour kit to become sensitive to these aromatic contrasts, it is central and not incidental to the process of learning to be affected. Latour argues that, 'The main advantage of the word 'articulation' is not its somewhat ambiguous connection with language and sophistication, but its ability to take on board the *artificial* and *material* components allowing one to progressively have a body' (2004: 210).

The concept of *articulation* moves the discussion of the senses and the sensual body beyond a singular body. As we have seen so far, work within body theory is reformulating the senses as forms of bodily awareness which are central to our movement through the world. This movement is one in which the senses cannot be isolated from more material and social and cultural processes. Latour's approach takes this further to explore the sensual body as one which cannot be studied by taking the singular body (albeit within a particular social and historical situation) as its focus. The body, for Latour, is an *assemblage* through the way it is connected up to material practices, human and non-human, which articulate its potentiality. The body is never singular and is always 'multiple' and in a process of becoming. We will explore this concept of the body as an *assemblage* in Chapter 5, which will focus on process, practice and multiplicity. What this brief introduction to the concept of assemblage and articulation draws our attention to is the idea that the odorous body, or the body capable of smelling, is always made possible by its conjoining and aligning with broader techniques and practices. Smell, therefore, is not simply an olfactory sense that makes the body more permeable and porous, but rather a potential 'capacity' of the body made possible by its *articulation* with broader techniques, artefacts and practices, such as the odour kit. This bodily capacity produces the body or bodies through a more dynamic definition that foregrounds the concept of 'learning to be affected' as that which ultimately characterizes bodily, sensual awareness. It recognizes that 'smell', for example, is a sense that is not biological or cultural, but a complex production and entangling of material and cultural processes.

HEALTHISM AND THE BODY

In this section we will consider an established area of body theory that considers the concept of the 'lived body' in the context of health and illness. The origins of this area started with a subdiscipline of sociology known as the 'sociology of health and illness'. The focus of work within this area of study took our lived, subjective experience of health and illness as its starting point. This was a response to what was taken to be the medical objectification of the body through the discourse and practices of clinical medicine. This objectification (the reduction of bodily experiences to signifiers of disease and illness) was seen to deny people agency and position them as medical objects. What was seen to be elided in this process was the lived experience of the illness, which might include how people were managing illness in the context of their own lives and the kinds of emotional or affective experiences produced by this. The late photographer and feminist Jo Spence (1995) documents such subjective, felt qualitative experiences in her book *Cultural Sniping*. This book combines personal autobiographical narratives with staged photographs representing what she felt was being ignored or left out by the practices of clinical medicine. Jo Spence was diagnosed with breast cancer and made a decision to undertake alternative healthcare treatments in the management of her illness. This was seen as an act of stupidity and ignorance by the medical practitioners caring for her. She also writes at great length in the book about how her emotional experience of disease, such as vulnerability and fragility, was ignored or elided.

This book was written in a context in the 1980s when, many sociologists have argued, doctors were fairly coercive figures of domination positioning subjects in and through the codes and practices of clinical medicine. The position of clinical practitioners met much resistance from academics, alternative practitioners and medical subjects who sought to refuse such domination. This refusal and resistance of the codes and practices of clinical medicine has become known as the 'medicalization critique'. This resistance to medical knowledge and advice might have included direct refusal and attack, non-cooperation, silence, avoidance and concealment (Bloor and McIntosh 1990). Sociologists have argued that the culture of health and illness has undergone rapid change and transformation. The body is no longer a body represented as a set of symptoms and there is more attention by medical practitioners to the context and lived experience of health and illness. Indeed, it is argued that the concepts of health and illness have so far undergone transformation that we need new concepts to understand the body in the context of illness and disease. One of the key concepts that will be found in the literature surrounding the body in health and illness is *healthism*. Healthism or 'self-health' is a concept that structures the new normalizing ways in which bodies are made intelligible throughout clinical, scientific and cultural practices. It has been argued that individuals are increasingly required to take on more and more personal responsibility for the maintenance of their own health. In this context, the body is never simply healthy or sick, but rather has a potential to 'break down' which is aligned to the effectivity of an individual's own self-managing strategies.

Although, one might argue, it is of course beneficial to health for individuals to engage in exercise, eat healthily and minimize stresses, the idea that health can be managed and regulated through the decisions and choices one makes places the burden of health management firmly within the hands of the individual. For this reason, the new culture of healthism is one that many sociologists view as a rather insidious strategy of self and social regulation. This is especially so in the context of the new threats to health and well-being whereby health risks to the individual become more uncertain and unpredictable. These new 'cultures of risk' include major epidemiological changes throughout populations, such as the global spread of HIV and AIDS; concern with food manufacturing processes; the use of additives; genetically modified food; the dangers of interspecies disease such as BSE and CJD, the dangers and possible effects of xeno-transplantation; the threat and risk of global pandemics such as avian flu and the global threat of terrorism and biological and chemical warfare. The emergence of these new cultures of risk have created a new set of problems for governments and a perceived need for a more intensive need for the microsurveillance and disciplining of populations. Healthism, in this context, could be viewed as a subtle and systemic form of management and regulation where individuals are required to take on such unpredictable risks through becoming more self-managing and self-disciplining. This is also known as 'hyper-individualism' and moves the focus away from macro-practices of government to the micro-practices of individuals. As Deborah Lupton argues, 'Central to this new emphasis on self-discipline is a focus between the imperatives of bodily management expressed at the institutional level and ways that individuals engage in the conduct of everyday life' (1997: 103).

SELF-HEALTH

One of the areas that has been explored in the context of 'healthism', and the new 'body cultures' related to the restructuring of health and illness in neo-liberal cultures, is the lived experience of receiving a cancer diagnosis. This work uses an approach to 'autobiography' that attempts to tie the lived experience of illness to broader scientific and cultural narratives; one such narrative is healthism, 'self-health'. Jackie Stacey's (1997) study of her own lived experience of receiving a cancer diagnosis is told in her book *Teratologies: A Cultural Study of Cancer*. This study is an attempt to explore the complex links between the stories we might tell about how we feel (autobiographical narratives), our experience of the body's materiality, affects and sensations, and how these processes are always mediated and transformed by broader cultural and scientific narratives. In other words, there is never any one-way direct perception of the body; it is always mediated. The important focus is how to understand this synthesis without falling into the trap of dualistic or binary thinking. Stacey begins her story with some of the feelings that structured her lived experience of her own body following a cancer diagnosis. She describes these feelings through the concept of abjection, which, as we have seen, is used to refer to the feelings of revulsion, disgust and horror we might experience when the border and boundary between the inside and outside is threatened. Stacey uses the concept to explore her experience of her own body as deceiving her, through a consideration of what normalizing narratives she had been living in relation to her body prior to its diagnosis.

Stacey asks, what happens when the boundaries or borders which construct conceptions of the body are disturbed. What boundaries and borders are often disturbed in our experiences of ill-health that usually remain hidden and which silently structure our experiences of the body? Stacey concludes that one of the predominant narratives that had been silently structuring her own experience of her 'healthy' body was the body as having truth-telling capacities. We are used to this narrative through the study of non-verbal communication and the way in which the body is seen to reveal its truth through subtle signs and codes. The body is aligned to 'bare life', to a state of nature and authenticity available to be read through attention to a universal language of bodily communications. However, the

---

Case Study

Kylie Minogue: Laughter is the best medicine for cancer cure

The 'triumph over tragedy' narrative was culturally valorized in the discussion of Australian pop-singer Kylie Monogue's acceptance and management of her cancer diagnosis. Her sister, Dannii Minogue told the British magazine *Cosmopolitan* that she is convinced that laughter had helped in Kylie's cure of breast cancer. Although the prognosis had apparently always been good they both kept up a positive attitude by watching comedies and wearing silly clothes which helped them believe that everything would be fine.

paradox for Stacey was that photographs taken when she was seemingly well (and which were commented upon as such by friends and families) were actually taken when the cancer was taking hold in her body. The body was not on her side, and although appearing healthy was actually masking the 'truth' of its demise, which was not available for surface inspection. She argues that the shock of a diagnosis is that this view of the body is disturbed. The body is deceptive and not to be trusted. 'The body tells a new story and so demands a reinterpretation of recent life history. Is it no longer to be trusted? Why has it withheld such crucial evidence? Whose side is it on anyway?' (Stacey 1997: 5).

Stacey explores how the shock of a diagnosis like cancer disturbs this conception of the body as having truth-telling capacities and demands a fundamental re-narrativization and refiguration of the person's relationship to their body. She argues that this is why cancer is such an abject term, an unspeakable term, why it creates feelings of shame, disgust and horror, why it is often only referred to in euphemisms, such as 'the C word' ... or is preferably not spoken about at all. 'Why should cancer disturb the subject so profoundly on a psychic level? Perhaps because cancer deceives? It silently makes itself at home and waits. The body which appears healthy hides the imminent truth of its own mortality' (Stacey 1997: 73). When it is spoken about, she suggests, it is usually through particular kinds of narratives. She suggests that these culturally valorized narratives appear in cultural artefacts such as magazines, autobiographies, literature, on television and in films. Closely aligned to the concept of healthism or self-health is one that Stacey terms the 'triumph over tragedy' narrative. This narrative presents the cancer sufferer as a hero rather than a victim, overcoming their diagnosis through positive thinking or accepting their fate with dignity and humility. They never fall apart, break down or get angry.

Kylie Minogue also carried out a world exclusive interview with Sky One in which she shared her experiences as a young woman diagnosed with breast cancer and the effects it had had on her and her family. She shared that she believed staying positive and believing you can get through is key to recovery and management.

NARRATIVES AND BODILY MATTERS

Stacey explores why the 'triumph over tragedy' narrative structures many auto-biographical accounts of living with a cancer diagnosis. She aligns its potency to Hollywood film narratives, which, she argues, tend to be structured through a very particular plot development and resolution. She concentrates particularly on the genre of the Hollywood action movie, which she says repeats endlessly a particular narrative structure. In action movies from *Independence Day, All in a Days Work* through to *Die Hard* national or even global security is threatened in some way. The threat is eradicated and calm restored usually by the actions of a particular hero who restores order and eliminates fear. A good example of the action genre that has been successfully extended to television drama is the multiple series of the American drama *24*, which stars Jack Bauer (Kiefer Sutherland) as the hero who has almost magical or superhuman powers to fight and eliminate the enemy. He never breaks down or dies, but carries on in conditions of extreme adversity, and often in the face of overwhelming pressure to give up or accept one's fate or lot. Stacey argues that this is a very omnipotent fantasy based upon a masculinist notion of control: that through one's own actions one can control the world and master evil, threat and even national and global disaster. Stacey suggests that this is a version of the 'triumph over tragedy' narrative that has become culturally authorized in discussions of the individual's management of disease and illness. We can see how this narrative, and its filmic dimensions, became fused in the constitution of the actor Christopher Reed as the living embodiment of his 'superman' role following his tragic horse-riding accident and consequent quadriplegic paralysis.

This is not to say that 'positive thinking' or hope and optimism are not important in the management of ill-health, but to question what other narratives are silenced or covered over? What experiences is the individual personally required to silence and omit through their own self-managing strategies? How does failure register? In other words, what happens if the person is not able to successfully manage their illness? Can this register as anything other than personal inadequacy or moral culpability? Stacey argues that these are the kinds of cancer stories which circulate and are validated within the media landscape. She is interested both in how these narratives structure personal accounts but also what they do not tell. When, where and through what narratives can feelings be acknowledged other than as personal inadequacy and failure?

Conclusion

This review of some of the key concepts that structure accounts of the 'lived body' has made visible some of the assumptions that structure this emerging area of body theory. The first is that studies are guided by a view of the body that is more 'experientially grounded' (Williams and Bendelow 1998: 8). These have also been described as more 'bottom-up' approaches that start with lived felt experience and then attempt to tie the production of our lived materiality to broader cultural and scientific narratives and practices. The concept of *embodiment* is often invoked as it refuses the idea that the biological and cultural, individual and social, the mind and body, for example, are separate entities that somehow 'interact'. This model of social influence, which we explored in Chapter 2, is replaced by various concepts, such as 'becoming' and 'learning to be affected', that view the body as an interface that is never singular and always tied or aligned to broader practices, entities and processes. To this extent the body is presented as a biologically and socially 'unfinished entity' (Williams and Bendelow 1998) that is not static or fixed. As Williams and Bendelow argue, 'mind and body, subjectivity and materiality are not in fact split, but are instead thoroughly intertwined' (1998: 54). We have seen how the *cogito* (the idea that reason and rationality housed in a separate mind characterize humanness) has been replaced by the sentient 'thinking' body. We have also seen how the sentient 'thinking' body is defined through its potential and capacity to be open to being affected and affecting. All of these approaches to the 'lived body' offer a reformulation of biology or the materiality of the body in non-reductionist terms. In this sense there is seen to be no *natural* body, but rather the materiality of the body is presented as a potentiality that is dynamic and open to being affected and affecting. The 'biological' body, or the body's materiality, is not therefore an 'autonomous physiological state' (Littlewood 1996: 15), but rather has a generative force that is not static or fixed. We will explore the implications of some of the insights of work on the 'lived body' in Chapter 5, where we extend this view of the body to include a focus upon *practice, assemblages, multiplicity and enactment*. The chapter will start with the premise that we cannot speak of '*the* biological body or *the* body in biomedicine as this body is not one (body)' (Davis 1997: 68).

5 THE BODY AS ENACTMENT

> I suggest that to think the body in relation to the senses is to: (1) encourage a thinking of the body in movement; (2) engage with the possibility that bodies are not limited to their organs; (3) shift the question of 'what the body *is*' to 'what can a body *do*'.
>
> Manning, *Politics of Touch: Sense, Movement, Sovereignty*

INTRODUCTION

This chapter will focus upon approaches to bodies that start from the position that bodies are never singular. Bodies are considered open systems that connect to others, human and non-human, so that they are always unfinished and in a process of *becoming*. We have explored the concept of becoming, which is a concept that moves beyond seeing bodies as fixed and closed to explore how they are produced and performed in specific ways, in Chapters 2 and 3. This chapter will focus particularly on the usefulness of the concepts of *multiplicity, movement, articulation, process* and *enactment* for understanding the production of bodies across different sites, locations and practices. None of the perspectives reviewed in this chapter ask what bodies are but rather focus attention on what bodies can *do*, and particularly on how they are *done*. The focus on *doing* rather than *having* or *being* a body is related to the concept of *enactment* which has been developed in body theory by anthropologists, sociologists, critical psychologists and cultural theorists who are working across the borders of their respective disciplines. The idea of border or boundary crossing is central to much of this work, which takes objects that are often the province of the natural, biological and life sciences and offer accounts of materiality, corporeality and the somatic (see Chapter 1) that are groundbreaking and challenging. They undo in different ways many of the assumptions that surround the myth of the natural body that has been a central focus of body studies more generally. Some of this work also moves beyond the human body in its focus and looks at the coupling of the human body with technology, nature, machines, animals and spirits, for example, producing new concepts for thinking the *doing* and even *undoing* of bodies.

PROCESS

The key concept elaborated in this section is *process*, which is distinguished from the body as *substance* or singular, bounded entity. In Chapter 4 we explored the way in which a French sociologist, Bruno Latour (2004), approached the sense of smell by considering the way in which the body of the perfumer becomes connected to various artefacts and techniques which allow him or her to finely discriminate smells. One of the techniques that Latour explores is the 'odour kit', which, he argues, allows or enables the body to become more and more aware of subtle nuances in smell. Latour draws on the concept of *articulation*, which he takes from the writings of the nineteenth-century American philosopher and psychologist William James (1902). William James did not see 'the self' as a fixed or self-enclosed bounded entity and explored the potential universe of becoming a self through his work on the 'subliminal self'. The term 'subliminal' refers to experiences that tend to exist outside of our conscious perception, much like the Aha experience that we explored in the Introduction. This is when we work through a problem without being consciously aware we are doing so (perhaps even whilst sleeping), and the solution might suddenly come to us 'in a flash'. The 'subliminal self' for James was the horizon of possibilities that could be actualized but exist in the background of our thought processes. Some possibilities are realized and others remain as an excess. This was captured by James with his notion of a 'stream of consciousness'. This describes the continual flow of ideas, affects, feelings, beliefs, memories and perceptions through our consciousness even though we might not be aware of, explicitly focus on or attend to this stream. This notion of a continual 'stream of consciousness' recognizes the multiple possibilities of *becoming* a self, or possible selves that potentially could be actualized or realized. The self for James was a self that was connected and permeable to this 'outside' of possibilities meaning that it could never be thought of as interior and closed (in contrast to the individualized, autonomous self, for example). Thus, to be *articulated* is to be open to connection, thus increasing the potential of bodies to be moved and to learn to be affected. In this formulation, learning is not a cognitive skill developed and undertaken by a brain or mind, but rather denotes the capacity of bodies to acquire more and more connections to artefacts, techniques and practices. It is the conjoining or coupling of bodies with practices and techniques that allow for what we might understand in this context to be their cognitive development.

The focus in this work on bodies-in-process rather than the body as a stable entity points towards the *multiplicity* and *movement* that characterizes materiality or corporeality. Maxine Sheets-Johnston argues that the corporeal turn across the humanities (that is, the turn to the body and body theory) should be comprised of a particular kind of corporeal turn: that we should 'be mindful of movement' (1999: xviii). For Sheets-Johnston, consciousness is always a corporeal or kinesthetic (see Chapter 3) consciousness that is created through the movements of singular and multiple bodies through space and time. Therefore, to 'think the body' requires a 'thinking in movement'. As she suggests, 'thinking and movement are not separate happenings but are aspects of a kinetic bodily logos attuned to an evolving dynamic situation' (Sheets-Johnston 1999: xxxi). Bodies do not remain fixed or static but are mediated by processes and practices that produce dynamic points of intersection and connection. The emphasis of work on bodies-in-process is *not* the body composed of particular parts, organs or entities. This is what Nikolas Rose (2007) terms a *molar* view of the body. The use of the term *molar* refers to some of the conventional ways we might refer to our own bodies, as being composed or made up of tissues, bones, limbs, blood, hormones and so on. We might then attempt to modify aspects of these entities through diet and exercise, for example. The focus on process is on *composing* rather than *composed, pre-formed entities*. The focus on composing looks at how bodies become assembled in particular ways through their coupling or conjoining with particular objects, practices, techniques and artefacts such that they are always bodies in the making rather than being ready-made. Let us take another example of this processual view of the body by considering a particular body/technology assemblage, that is, the body as it might be enacted or performed through dance. The following section will consider how we might approach dance through a focus on bodies-in-process, or what I have also called the *composing* rather than *composed* body. We will consider recent work that has engaged with the dance known as Argentine Tango through this model.

THE BODY-IN-MOVEMENT

Tango is evoked through a politics of touch that resides in the intent listening to(ward) an other. This attention to a gesture carried within the movements of the body is a listening that carves space in time with our sensing bodies in movement. In the best cases, there is not one dance to be danced, but a myriad of possibilities generated by two bodies, often foreign to one another, touching one another. I lead, you follow, yet even as I lead, I follow your response, intrigued by the manner in which we interpret one another, surprised at the intentness with which our bodies respond to each other. (Manning 2007: 17)

We might think of learning to dance as a kind of apprenticeship, much like becoming a perfumer, that demands a particular kind of knowledge and learning. We might be prepared to think of the body as a malleable entity in the sense that we are open to seeing changes through our learning to dance, perhaps in our posture,

breathing, musculature and body language, for example. We might also be prepared
to accept that learning to dance is not about the isolated, singular, molar body
but requires a conjoining with others, human and non-human. This might be a
partner, a specific pair of shoes, a designated dance space, a music system, a space
to change and use the toilet and so on. Some forms of dance, like ballet, require
the dancer to develop the capacity to ignore pain, hunger and exhaustion (Aalten
2007). Thus some forms of dance might be marked by the association of particular
states of being, such as exhaustion and hunger, with particular aesthetic shapes and
body forms, such as the light, slender, body of the female ballet dancer. This might
involve developing a relationship or orientation to the body in which one views the
body as an instrumental machine that can be denied food or where pain and injury
can be overridden. As Aalten argues, 'the ability to control one's appetite and to go
without food in order to reach the ideal of the disembodied woman was all part of
the socialization of the dancer' (2007: 118). This approach to dance, characteristic of
work within the sociology of the body which we explored in Chapter 1, focuses upon
the cultural practices and body techniques that allow particular kinds of corporeal
transformation. As Nick Crossley, argues, 'the concept of body techniques poses the
question of the evolution of particular uses of the body' (2007: 92). That is, dance as
a particular set of body techniques and practices can tell us about how bodies can be
modified, altered and transformed.

However, a focus solely on body techniques and practices, Crossley suggests, misses
out an important component of the composing body or bodies-in-process. That is,
the more sensual and felt components of learning to be affected and *becoming*, in
this context, a dancer. This might involve developing a bodily sensitivity or openness
to connection that cannot be found in manuals or taught by instruction, command
or even example. This is the focus of a recent book by Erin Manning (2007), *Politics
of Touch: Sense, Movement, Sovereignty*, that considers what the body *does* when it
learns Argentine Tango. The book focuses on how bodies change, alter and transform
not just through acting upon their physicality through body techniques, but how
they are modified and transformed 'as a result of touch' (Manning 2007: xi). The
conception of touch that Manning develops does not view touch as a physical sense,
but is closer to the reformulations of touch that we explored in Chapter 4, through
the idea of skin knowledge. This is, a concept that refers to a different kind of bodily
knowing or awareness that moves beyond seeing touch as a literal, brute, physical
sense (Howes 2005; see Chapter 4). The version of touch that Manning develops
relates to a 'sensing body' that is always in movement. Touch, Manning holds, is a
relational sense. Touch connects us to others and is also a register through which we
are articulated with others. Manning uses the concept of articulation, in a similar
way to Bruno Latour, to refer to all those possible relational connections that exist

and which change and alter bodies as they move and sense in the world. It is through this openness to possible relational connections that worlds are created and bodies become. In this view, we are not talking about the coupling of a stable, preformed body with another, human or non-human, but rather the body as a process that is continually in the making. The body is always co-constituted through its relations with others, human and non-human.

Manning draws on a range of concepts found in the philosophy of Deleuze and Guattari (1987), Simondon (1992) and Brian Massumi (2002a and 2002b) to develop Argentine Tango as a figuration for thinking through the *processual body*. The concept of *figuration* comes from the writings of the feminist philosopher Rosi Braidotti (2002) and the feminist science studies writer Donna Haraway (2004). A figuration is the act of forming something into a particular shape and is usually used to refer to the pattern, form or outline that occurs as the endpoint of this process. Haraway uses the concept of figuration as a heuristic device or thinking tool for drawing attention to some of the patterns and repetitions that characterize what she terms our inherited thinking on particular subjects. These might be the idea that we are separate, bounded individuals who can be clearly delineated from others: machine and animal, for example. For this, Haraway (1991) mobilized the *cyborg* as a figuration for moving beyond the idea of separation between human and machine. The cyborg is part machine, part human, a strangle coupling in which neither human nor machine can be differentiated or finally settled. Manning uses the concept of Argentine Tango as a figuration for making visible some of our inherited assumptions about the body and the senses, which the dance challenges in the way it is taught and experienced. These are that the senses can be located within a singular, molar, bounded body, and that bodies are static, pre-formed entities. She argues that Argentine Tango discloses the ways in which bodies are always in excess of themselves, and therefore *multiple*. As she says, 'there is more than one way for a body to become' (Manning 2007: xx). Touch discloses these potentialities for multiple body-assemblages and Argentine Tango provides an interesting figuration because it is based upon improvisation and, arguably, can only be learnt through sensing the other and the profound relational connection that ties you together in the dance.

BODIES WITHOUT ORGANS

One of the concepts central to Manning's work, and to that of many other scholars interested in the processual body, is the concept of BwOs (Bodies without Organs) derived from the philosophical writings of Gilles Deleuze and Felix Guattari (1987). This section will explain this important concept and relate it to the concept of articulation, which we have already explored in this chapter. The concept of BwOs encapsulates an attempt to go beyond seeing the body as a molar entity or organism (made up of blood, organs, limbs, hormones etc.), and rather to see the body as always extending beyond itself and being conjoined with or articulated by practices, technologies, institutions, objects and so on. It is about the relational connections or *networks* that produce bodies as very particular kinds. However, the importance of the concept of BwOs is that the production of bodies is never fixed and is marked by the possibility of mutation, transformation and change. This is where the importance of the concept of *movement* shifts the terrain of simply seeing bodies as discursively or socially constructed. Bodies are not simply stabilized effects of subject-positions that precede them (see Chapter 3). This, according to Brian Massumi (2002a and b), freeze-frames the subject providing a limited and very particular snapshot in time. As Manning argues, 'bodies can be stratified, organized, categorized, even restrained, but they cannot be stopped' (2007: 135). The body is always *with* and defined by the relational connections that can be sensed and which articulate its sensing at particular moments of lived time. Although particular body assemblages might have a relative stability they are always open to change and movement. The skin is therefore not seen to be an enclosure or container of the self but rather an interface meaning that the self is always gesturing beyond itself. We saw a similar view of the skin in Chapter 4 where we explored the concept of the *skin ego* in relation to the writings of the French psychoanalyst, Didier Anzieu (1989). Again, as Manning suggests, 'the BwO is a way of asking ourselves how we could ever have thought of the body as a unit, as a uniform singularity' (2007: 136). The body is therefore, within this perspective, always in movement and always multiple. For Deleuze and Guattari (1987) and those that have been inspired by this processual view of the body, bodies can never be mapped on to or overlaid by the organism or molar body. The body is always much more than its organs, limbs, blood and hormones, for example. As Braidotti (2002: 21) suggests, the body has a materiality which is 'not of the natural, biological kind'.

In order to provide some context to the development of BwOs it is worth spending a bit of time outlining the ethics and politics of Deleuze and Guattari's philosophy and their concept of becoming, which is intimately related to BwOs. In Chapter 3 we saw how the Deleuzian concept of becoming has been taken up and developed by feminist philosophers interested in the processual body, such as Elizabeth Grosz (1994) and Rosi Braidotti (2002). This work draws attention to the multiple possibilities of becoming that exist even in the face of the normalizing processes we explored in this chapter. Although certain bodies might be produced, through the reproduction of norms, as legitimate and 'bodies that matter' in the words of Judith Butler (1993), there are always possibilities to 'fight back': to become other. The notion of *becoming other* is related to Deleuze and Guattari's concept of *becoming woman*. Rosi Braidotti (2002) suggests that the concept of woman is 'molar': it is produced as an entity or subject position through normalization but cannot capture or contain the embodied specificities of becoming woman in all its possible multiplicity. Becoming woman relates to what Guattari (1984) understood as *molecular* desire: that which is about the movement and flux of desire that potentially creates diversion, distraction and new ways of seeing, being, experiencing and inventing bodies. Molar desire is related to *territorialization* (that which becomes institutionalized, stabilized and made to circulate as truth), whilst *de-territorilization* is equated to the movement of molecular desire that always produces the possibility of change and transformation rather than fixity. Braidotti (2002: 70) uses the figuration of 'nomadic subjectivity' to refer to the transience, mobility and changeability taken to define bodies-in-process. *Subjectivity* is a concept mobilized by Braidotti to refer to the ways in which subjects attempt to create a coherence or what she terms 'fictional unity' (2007: 75) for themselves in the face of these contradictions and multiple possibilities. Braidotti (2002) develops this argument in relation to *affect*, which will form the discussion and basis of the Conclusion. In this section I want to finish by considering whether the concepts of BwOs and becoming other are romantic and even utopian.

The concept of BwOs and becoming were, always for Deleuze and Guattari, a political vision for change and transformation that might be possible but were not necessarily actual or actualized. On the one hand, both concepts draw attention to our intimate *mixing* and interconnection with others, human and non-human. These might be animals, insects, spirits, machines and so forth. Both concepts draw attention to multiplicity, plurality and 'not-oneness' (Braidotti 2002: 114). However, Braidotti also considers how 'mixing' or interconnection tends to be made to signify within advanced liberal cultures. She draws attention to popular and literary representations of mixings of animals with humans and how these representations tend to be marked by horror, disgust and fear as well as fascination. She cites the iconic images of the vampire, werewolf and demon as those that stand 'simultaneously for ethnic mixing, moral ambiguity, sexual indeterminancy and unbridled erotic passion' (2002: 128). In a similar way, Donna Haraway (2004) has also written about mixing and interconnection, particularly of human and machine, in her figuration of the cyborg. This figuration draws attention to what she calls the 'stigmata of morphing' that often accompanies the blurring of such boundaries (2004: 281). She cites the example of Michael Jackson, the American musician known not only for his innovative disco music and dancing but also for his participation in practices such as cosmetic surgery. She argues that 'beginning unambiguously as an African American boy with striking talent, Jackson became neither black nor white, male nor female, man nor woman, old nor young, human nor animal' (Haraway 2004:

281). However, this blurring and in-betweenness is a much less safe representation than the panhuman, multicultural 'mixed up' images of Benetton, the global fashion brand who have used images of racialized mixing in their own fashion advertisements where such racial blurring is presented as stylish, fashionable and beautiful (see Franklin, Lury and Stacey 2000). We might also think of popular, literary and filmic representations of twins and twinning as another example of multiplicity and mixing. We could examine whether the special relationships that twins might have is also marked by horror, disgust, fear and revulsion as well as fascination and attraction? This might include the common perception of twins or twinning as being marked by a pathological closeness or excessive dependence, for example.

When we take into account how *interconnection*, *multiplicity*, *in-betweenness* and *not-oneness*, are often represented as objects of fear as well as fascination, we can see how the concept of BwOs could also be seen as utopian and a slightly romanticized version of what it means to enact our subjectivities. These are debates that can be found in the literature and which are far from settled. It would seem that to invoke the concept of bodies-in-process inevitably results in discussions about subjectivity, about how we 'hang together' and develop a coherent sense of self. We saw in Chapter 2 how in advanced liberal cultures the 'individualized' clearly bounded and

Case Study

The anxiety over cloning is actually part of a much older fear – one that ties in with the motif of the doppelganger or double. This is the fear that our essence or soul may be stolen, captured or transferred to our double, who competes with us for the right to live. Individual identity is therefore threatened by the doubling, dividing or interchanging of the self. (Nottingham 2000: 53)

Nottingham suggests that the trope of twinning provides a repository for our fears and anxieties in relation to the idea of doubling or multiplicity which are being reactivated in the present through cloning. Cloning is a technique which can produce copies or replicas of animals (or potentially even humans) through copying

genetic material from the cell of one animal into an egg from which the nucleus has been removed and implanting the reconstructed egg in the womb of a host mother. Cloning and twinning both capture the fear and fascination that multiplicity presents to our individualized conception of selfhood. These fears and fascinations have been the subject of numerous films that use the motif of the double to explore our anxieties about a single entity becoming multiple or double. Nottingham surveys a range of films which use the motif of the double to explore these fears and fascinations. These include *The Stepford Wives* (1974) in which humans (women) are replaced by dumbed down versions of themselves which will perform certain tasks (cooking, cleaning, sexual acts) but not think for themselves or challenge their husbands. This reflects separate self, what Nikolas Rose (1989) terms, the 'fiction of autonomous selfhood' has become part and parcel of how we are governed and managed as citizens and populations. This version of selfhood endlessly circulates so that it has taken on a 'truth-value' or status and has become part of how people judge, evaluate and act upon themselves as subjects. This 'molar' view of the self is one that easily maps on to the 'molar' view of the body that proponents of the processual body are keen to dismantle and displace. Rose (2007) suggests that although this 'molar' view of the body is related to a nineteenth-century clinical gaze (see Foucault 1973) biomedicine increasingly visualizes the body through a molecular gaze.

THE MOLECULAR BODY

How did we become neurochemical selves? How did we come to think about our sadness as a condition called 'depression' caused by a chemical imbalance in the brain and amenable to treatment by drugs that could 'rebalance' these chemicals?

Rose, 'Becoming Neurochemical Selves'

a fear of how the 'self' might be taken over by an imperfect or inferior copy or replica and be engineered so it can only perform certain roles or functions.

Nottingham suggests that clones and twins within films are often portrayed as damaged, as having reduced intelligence or having physical imperfections. Roles in which opposing characteristics are represented by the good and evil twin are frequent. These polarized characteristics are often aligned to crime, mental imbalance or even schizophrenia; twins are presented as the pathological 'Other' to the separate, clearly individuated self. We can see this as a theme in David Cronenberg's (1988) film Deadringers in which the evil twin commits murder in order to resolve a longing he has for a woman with whom both he and his 'good' twin have fallen in love. The theme of cloning has been used as a device in a range of films where the clone is usually represented as inferior, and subtle distinctions are used to differentiate the clone from their original. This includes the 'android' in Bladerunner and the cloning of aliens in films such as the Alien Resurrection. Both films play on our desire to differentiate the clone from the original, and our concern over what might happen if a clone could love or show feeling for example. This was also explored in Stephen Spielberg's film Artificial Intelligence. We can clearly see when twins and cloning 'go to the movies' they are mobilized to play on our fears, fascinations, repulsions and desires, and thus remind us of how normative the notion of separation and individualization is to what Rose (1989) terms the 'fiction of autonomous selfhood'.

The molecular gaze is related to a view of the human body as not being a closed, living, functional system but an open system that combines with other systems so that it always extends beyond itself. The body is not made up of distinct parts, within this view, but processes (that Rose suggests occur at the level of code, ion channels, enzyme activities and transporter genes) that can be altered, modified and manipulated without needing to be housed or contained within the molar body. The new biotechnological techniques that can intervene at this molecular level enable these processes to be 'de-localized' thus creating a new mobile conception of life (see Rose 2007). He argues that we have become '*somatic individuals*' increasingly understanding and acting upon ourselves through this new language of biomedicine, seeing personhood as being open to modification and alteration at the molecular level. Rose asks how it is that we have come to understand our selfhood or subjectivity in terms of our brains and particular biological understandings of the body (albeit located at the molecular level of life). The idea that depression has a molecular biochemical basis is central to biomedical understanding and also permeates popular culture. On the basis of these understandings new so-called smart drugs have been developed that will target particular biochemical pathways and restore us to some version of normality. Rose points to the World Health Organization's proclamation that 'mental illness is a treatable brain disorder treated with medication just like diabetes is treated with insulin'. (2007: 105). One such smart drug is Prozac, one of the first SSRIs (Selective Serotonin Re-uptake Inhibitors) to have been marketed by drug companies as having a specificity in terms of its action on serotonin, an imbalance of which is seen within the biomedical culture to be responsible for depression. As Rose suggests, it is difficult to think outside of this discourse, and this is no more so than within psychiatric practice that reduces the body to brain and mental distress to brain pathology. Mental ill-health is seen to be 'residing within the individual brain and its processes' (2007: 122).

Case Study

One example of the molecular body can be found in discussions of the rise and role of anti-depressant drugs, known as SSRIs (selective serotonin re-uptake inhibitors). The most popular of these drugs goes under the brand name of Prozac, the ubiquitous use of which is held by some not simply to normalize depression but rather to be capable of actively re-engineering and enhancing personality at the molecular level. The debate is about whether these drugs are confronting the molar view of the body as a fixed entity or presenting

Drugs such as Prozac have also been marketed through particular cultural logics. As Metzl (2003) argues, Prozac has not simply been presented as a 'smart' drug (as having a biochemical specificity, for example), but has also been marketed as enabling the user to become 'optimistic, decisive, quick of thought, charismatic, energetic and confident' (ibid.: 15). The depiction of a user following an encounter with Prozac is of 'a generative working member of society, holding fruitful employment in her productive days'. Biology is thus *socialized* and given meaning according to prevailing conceptions of normative personhood (see Chapter 3). Metzl suggests therefore that drugs perform: they enact particular norms where their potency extends far beyond any specific action on brain processes. Prozac, for instance, is primarily seen to enact or perform a 'productivity narrative' (2003: 173). In this sense, as Emily Martin (2004) argues, everybody has the chemical potential to achieve such norms and above all to 'feel good'. Fraser (2001: 59) maintains that Prozac is therefore an 'enhancement technology' which is presented as making people 'better than well' (Kramer 1994 quoted in Fraser 2001: 59). The key question for consideration is the extent to which these drugs and the cultural norms they perform are making us become 'more biological', or the extent to which the drugs present the self as fluid and flexible, and therefore subject to change. Another way of framing this discussion is to ask whether the drugs merely 'cure' or whether they enable the reshaping of personality at the molecular level. If it is the latter, then what kinds of cultural norms are desirable, particularly if personality can be said to be mutable rather than fixed and static? To what extent might this be reducing the body to a synthetic medium or code that disavows or avoids the complexity of the relationships between bodies, affect and culture? These are questions for classroom discussion that are taken up at the end of the book.

---

the body (as a set of molecular processes) that can be modified, changed and transformed. The question becomes one of whether this challenges the view of personhood organized around the 'fiction of autonomous selfhood' or whether, rather than becoming-other (as this challenge might suggest), we are potentially 'becoming more biological' (see Rose 2007). You can consider this debate for yourself by reflecting upon how you might understand the increased consumption of anti-depressants in industrialized societies.

POSTHUMAN BODIES

Ours is a world where cloned calves and sheep carry human genes, human embryo cells are merged with enucleated cows' eggs, monkeys and rabbits are bred with jellyfish DNA, a surrogate horse gives birth to a zebra, a dairy cow spawns an endangered gaur, and tiger cubs emerge from the womb of an ordinary housecat.

Best and Kellner, 'Biotechnology, Ethics and the Politics of Cloning'

The social representations and fears related to twinning that this might entail persists in contemporary representations of biotechnological practices that blur the boundaries between the natural and the technologically manipulated or altered body. The techniques of cloning are perhaps the closest to the contemporary fears and fascination surrounding twinning as they provide the possibility of creating doubles or multiples, human or animal, that blur these very distinctions. These techniques, such as transgenic genetic engineering, blur the boundaries between the natural and the technological. Transgenic engineering involve the insertion of a gene from one species into another 'to suit the needs of science and industry' (Best and Kellner 2004: 56). Cloning involves producing identical replications or multiple copies species bodies via the processes of gene technology, with the potential to be utilized 'as organ sources for human transplantation' (2004: 58). The genetically new kinds of bodies created through these practices put into question, as Joanna Zylinska argues, 'the origin and nature of 'life itself'' (2005: 140). Although those that have come to the attention of the popular imagination include Dolly the Sheep, a 'sentient transgenic being' (Zylinska 2005: 143; also see Franklin 2007), they have also become iconic figures of transgression that have become the subject of bioethical debates about how far 'nature' should be altered, modified and transformed. One problem has been the failure of these practices thus far and the attendant suffering and deformed monstrosities that have been produced through such practices. As Best and Kellner argue, 'even Dolly became inexplicably overweight and arthritic, and may have been prematurely ageing. In February 2003, suffering from progressive lung disease, poor Dolly was euthanized by her 'creators', bringing to a premature end the first experiment with adult animal cloning and raising questions concerning its ethics' (2004: 60).

These debates about how different species bodies are being *done* or enacted within biotechnological practices relate to questions about the status and basis of *life*, *nature* and *humanness* and how these unstable entities connect to understandings of *power*, *subjectivity* and *technology*. These questions have become central to discussions about the *posthuman* which primarily emerged as a category of thinking in relation to the perceived new relationships between technology and bodies that were emerging at the turn of the last century. The posthuman refers to the destabilization and unsettling of boundaries between human and machine, nature and culture, and mind and body that digital and biotechnologies are seen to be engendering. Zylinska discusses the practices of the digital artist Stelarc, who has performed the conjoining of his body with technology (digital and biotechnological) in order to foreground what she terms the 'original prosthecity' of selfhood and its situatedness in the network of relations that criss-cross the envelope of the skin' (2005: 132). The idea of the body as having an 'original prosthecity' relates to the concept of the body-in-process whereby rather than the body being conceived as 'molar' it is viewed as a series of 'co-dependent additions and replacements' (Zylinska 2005: 132). It is never singular and bounded but rather governed by an *alterity*, a concept that comes out of various philosophical traditions that see the self as always relational, always defined by its interconnections with others. To be aware of this profound relationality is to be open to it, with the result that the body is always 'intrinsically other' (Zylinska 2005: 123) or 'beside-itself' (Butler 2004). This includes, within Zylinska's perspective, being open to our profound relational connection to practices and techniques that enact or *do* the body in radically different ways. Stelarc and the French performance artist Orlan are both, according to Zylinska, are willing, capable and able to open 'their bodies to the intrusion of technology' (2005: 123). Orlan, for example, has utilized practices of cosmetic surgery to challenge normative conceptions of female beauty. This has included having a cheek implant placed under the skin of her forehead, and using digital technology to devise the biggest nose that her facial structure will be able to support. As we can see, being open to connectedness is marked by controversy and ethical dilemmas, as well as making visible our inherited thinking on the body.

The idea of 'prosthetic selfhood' (Lury 1998) can be extended to our relationships and interconnections with all manner of practices and technologies. The concept of prosthesis being mobilized here, is one, as we have seen, that is about our profound connection or relationality with others, human and non-human. It is not simply about the prosthetic (limb, teeth, glasses) as being replacements or additions to the human body augmenting and amplifying its capacities (usually framed through a loss of the original). Rather, it is a concept that blurs the boundaries between the human and machine, the natural and cultural and so forth. Katherine Hayles in her book *How We Became PostHuman* (1999) engages with some of the new kinds of molecular thinking about bodies that are dominating the life, biological and cognitive sciences. She argues that this molecular gaze reduces the body to code, to information, thus losing a conception of embodiment and the lived body (see Chapter 4). She argues that, 'a defining characteristic of the present cultural moment is the belief that information can circulate unchanged among different material substrates' (Hayles 1999: 1). This is the new apparent 'mobility of life' that Nikolas

Rose (2007) argues is a marked characteristic of some of the biotechnological practices that are challenging conceptions of life and nature. Hayles's argument rests upon the assumption that what is created if the materiality of bodies can be reduced to *information* is a *dis*embodied view of what it means to be human. Although the molecular view of the body is enacted within many biotechnological and digital practices, and challenges what she speaks of as 'the perceived boundaries of the body' (Hayles 1999: 23), one of the pressing considerations for body theorists is how to present the complexity of our fleshy materiality that takes into account some of the very many different issues we have been examining within the book so far. It would seem that we are far from merely being molecular bodies that can be reduced to code and information to the extent that the boundaries between machine and human appear unstable. Rather, she asks, 'how much had to be erased to arrive at such abstractions as bodiliness information?' (Hayles 1999: 12).

We can see, then, that although bodies are being *done* and *enacted* within bio-technological and some digital practices in ways that disrupt or disturb the boundaries between human and machine, we also need to be mindful of what might be being written out of these accounts of the corporeal, somatic and material. As Barbara Stafford (2007: 102) argues, the organism is being transformed by such practices into a 'synthetic medium'. What this does is replace the relationality and connectivity that defines life creating an 'oddly depersonalized and distributed conception of the contemporary body' (Stafford 2007: 108). If different practices articulate different kinds of bodies, then the rise of a very particular *synthetic* or *informational* body, within biomedicine, for example, creates a range of ethical dilemmas for contemporary body theorists. We will examine later in the chapter how we might approach the multiplicity of ways in which bodies are enacted across different practices, and the implications of this for how we might examine embodied subjectivity. Although the concept of the posthuman draws attention to the new kinds of relationships being invented with ourselves and others, particularly with machines and bio-technologies, and the new kinds of bodies becoming possible and probable, the concept is limited in its scope and vision. The body cannot be reduced to discrete molecular or informational units without erasing the complexity of the *lived body*, which is so *much more* than its biotechnological counterpart, the synthetic or molecular body.

COMPANION SPECIES

we have never been human, much less man

Haraway, *The Haraway Reader*

In the Introduction we considered an account written by an Australian sociologist, Ann Game (2001) of her relationship with her horse, KP, and the mutual connection, mixing and radical rationality that enabled KP to canter and trot following her paralysis. This moving account displaced the idea of a distinct separation between horse and human, and instead pointed towards the connectedness and openness to such connectedness which defined their relationship. She displaces the idea that their relationship was one of two discrete and separate entities, horse and human, interacting, to one of which she contends, 'there is no such thing as pure horse or pure human. The human body is not simply human' (Game 2001: 1). We also encountered another example, in Chapter 2, of a horse–human relationship which likewise dispelled the myth of a clear and distinct separation between human and animal. This was the example of Hans the Horse, the horse who appeared able to solve complex arithmetic problems by stamping his hooves. In order to understand how this might be possible, other than seeing Hans as a genius or a psychic, the concept of attunement was invoked, referring to subtle mixing processes, that often we might feel or sense but of which we are not consciously aware. It was argued that the horse and human might be subtly attuned to minimal body movements that enabled communication to take place beyond conversant awareness. This is closer to the sense of touch that Manning (2007) utilizes to explain the practice and experience of Argentine Tango as a dance that requires a 'sensitive feel' or awareness of connection. In this context, training, whether learning to ride or learning to dance, always involves mixing, human and non-human. In this section we will further consider the connecting and mixings between human and animal by examining the usefulness of Donna Haraway's (2003) figuration, which she terms *companion species*: 'There cannot be just one companion species; there have to be at least two to make one. It is in the syntax; it is in the flesh. Dogs are about the inescapable, contradictory story of relationships – co-constitutive relationships in which none of the partners pre-exist the relating, and the relating is never done once and for all' (Haraway 2004: 300).

It is a well-known adage that pets often look like their owners and vice versa. This is one popular version of the connectedness of humans with their pets that has almost taken on the status of a truism. A study in 2001 by an American psychologist confirmed that a relationship of similarity was more likely amongst pet-owners who had chosen a pure-bred dog as a companion. There is also a thriving consumer market of various dog apparel, fashion accessories, clothes, custom-made dog collars and jewellery and lotions for pampering your pooch. However, this version of connectedness, although it is based on the idea of the dog as a companion animal (which might also afford health benefits, for example) and recognizes how the pet and owner might merge is not reducible to Haraway's figuration of the companion-species, which is intended to make visible the co-evolution and co-constitution of dog with human and vice versa that troubles the clear distinction between nature (biological kind) and culture. The figuration of companion-species is designed to trouble the category of the separate, individualized self, and to think the relational connections between human and animal when they are not conceived as separate entities or 'hermetically sealed objects' (Haraway 2004: 303).

In the story or *version* (see Chapter 2) that Haraway tells, evolution is not the gradual one-way domestication of wild animals to the point where they become more civilized and able to live with humans in domestic settings. Rather, humans and animals are seen to have taken shape in their interactions with each other, and through the practices that have afforded certain connections to emerge and evolve. In other words, dogs and humans have enacted particular co-constitutive relationships, in particular settings (that include particular artefacts, technologies and instruments, pedagogies and training, for example), that have produced many different entities (dogs and humans) and many kinds of relationality. The companionship in the figuration is about the 'something else' (Haraway 2004: 331) that emerges when the dog and human are not considered 'pure' species. The capacity for 'dogness' might, therefore, be related to the capacity for being open to connection and alterity that could be enacted in a myriad of ways (mastery, control, possession, attunement, sensitive feel and so forth). It is the relational connections that are enacted that produce what the entities are taken to be and *become*. It is these mixings, connections and entanglements that trouble any easy distinction between nature and culture and are captured by her concept of *naturecultures*. The weaving together of two terms that are usually considered separate entities, nature and culture, is a gesture towards displacing the concept of social influence (individual/society dualism) which we explored in Chapter 2, and moving towards theories and concepts that can explore the co-constitutive processes that produce and enact bodies in all their diversity and materialities. The concept of separation is replaced with a radical relationality that explores the connections and linkages which make it possible to enact *multiple* bodies that do not end at the skin, and which are always oriented towards Otherness (alterity). In the next section we will explore a theory that attempts to provide a framework for exploring such linkages, and which maintains the proposition that there are no *essential* differences between human and non-human actors pre-existing their linking and co-constitution. This does not mean that the human and the non-human are the same, but that their differences are continually being made and re-made. One theory which mobilizes this view and which makes a useful contribution to body theory is known as Actor Network Theory (ANT).

ACTOR NETWORK THEORY

French sociologist, Bruno Latour (2005a), who is already known to you through his work on the body and the concept of *articulation* (see Chapter 4 and this chapter) is responsible for the development of ANT. One of the key assumptions of this theory, relevant to work in body studies, is that the concept of providing a 'social' explanation of the body as opposed to a 'natural' explanation is a misguided aim for scholars from the humanities interested in the body and body theory. The idea of simply adding a social explanation to a natural explanation should, according to Latour, be considered a problem, rather than a resource for body theory. In other words, work on the body and embodiment across perspectives from the humanities, as opposed to the natural, biological or even psychological sciences, should not be framed as simply 'adding' a social explanation to pre-existing biological or psychological explanations of our materialities. Although, as we have seen throughout the book, work on body theory does tend to unify around the problem of how to 'think the body' as a complex set of interconnecting psychological, social and physical processes, it recognizes that these processes are never pre-existing, unified entities that can be easily untangled. Bruno Latour goes slightly further than this in his formulations and argues, instead, that we are not simply providing accounts of the social dimensions of bodies but, rather, should be rejecting altogether the concept of the social as having an explanatory power or purchase. The problem for Latour, which is relevant for body theory, is the way 'the social' is often mobilized as a preformed entity to explain other entities, such as the body. He argues that 'there is no social dimension of any sort, no "social context", no distinct domain of reality to which the label "social" or "society" could be attributed; that no "social force" is available to "explain" the residual features other domains cannot account for' (Latour 2005a: 4)

If there is no 'social' that can be used to explain other entities, such as the body, then where does this leave body theory? As we saw in Chapter 2, one of the concepts that has been developed within different ways across body theory is an embodied paradigm, and the concepts of embodiment and the lived body. Both of these concepts attempt to move beyond the idea of the natural and cultural as being two separate entities which somehow interact, to explore their complex entanglement and interweaving (Csordas 1994). One of the problems with this work, however, is the way it tends to remain with the human subject and the way they narrate or experience their corporeality. In other words, the focus is on the 'intentionality' of the subject: the way they intend or make sense of their bodies, in health and illness, for example. This becomes more problematic when we want to consider the relationship between these narratives of experience and their interweaving with particular techniques and practices. As we saw in Chapter 2, the bringing together of the natural body with what is considered 'outside it' has been stabilized through the concept of social influence. Although the concept of social influence adds a 'social' explanation to what is conceived as a naturalistic body, it reproduces a split or dualism between the individual (intending subject) and the 'social' (as a separate entity). We explored the concept of social influence in different ways in Chapter 2 with our focus upon the 'communicating body', and how it differs from the concept of becoming. As we have seen in this chapter, the concept of becoming suggests an understanding of the body that relies upon connectedness and mixing rather than singularity and separation. Rather than ask in relation to the body, 'is it nature or culture?' (Despret 2004b: 35) we instead looked for other versions of the 'communicating body' that might trouble the idea of separation between individual and the social. This included exploring the phenomenon of emotional contagion and the placebo effect as examples that disrupt the concept of social influence (see Chapter 2).

In a similar way, Latour suggests that the use of the term 'social' homogenizes those elements and processes, human and non-human, that have been connected together to produce what he terms a 'well-formed *assemblage*' (2005a: 8), meaning those relational connections which are relatively stable. The term 'assemblage' is a reference to the assembled connections that produce and enact what entities, such as the body, are taken to be. Although assemblages are well-formed connections which might have a semi-permanence, the connections are always subject to modification, alteration and recomposition and are thus *temporary* in nature. One of the most important aspects of assemblages is that they are performative: they are an association and concatenation of a range of heterogeneous elements which *produce* what we take entities to be. These entities are themselves a mixture of processes, social and non-social, that cannot be disentangled as they are produced and enacted by the assemblage. In other words, if we use the concept of the social to describe bodies, we must be prepared to trace the social and non-social elements that have become tied together and distribute our bodily capacities and potentials in specific ways. The different distributions of the body-in-process *do* or *enact* the body in radically different ways, rather than simply reveal or disclose a naturalistic object – the body as *substance*. As Latour suggests, 'In most situations, we use 'social' to mean that which has already been assembled and acts as a whole, without being too picky on the precise nature of what has been gathered, bundled, and packaged together' (2005a: 43).

In this sense, the social cannot simply be 'added' to the biological or the physical/material, but is, rather, a term that describes the heterogeneous objects, human and non-human, which in their association and co-existence make it possible to *do* certain things. These heterogeneous objects are always, 'complicated, folded, multiple, complex and entangled' (Latour 2005a: 143). They never pre-exist in *kind* or substance the relational connections which produce and enact them as very particular types of object. In other words, it is the relational connections that articulate what the body is capable of, what it can *do*, what it might *become*. Although Latour (2004) gives the example of the perfumer's body in his article 'How to Talk About the Body? The Normative Dimensions of Science Studies', the two key scholars to have developed the implications of this work for body theory are John Law, the British sociologist, and Anne-Marie Mol, the Dutch anthropologist. Both scholars particularly focus upon the challenges of reimagining method (how we might study bodies) in light of these new concepts for thinking bodies. In order to illustrate what can be a very abstract theory and set of concepts, I am going to turn to one of their studies, which illustrates the specificity of ANT for approaching the body. This study takes the condition of hypoglycaemia as its object, and explores how hypoglycaemia is enacted and performed in a multiplicity of practices which produce it as very different kinds of object. The focus in this work, which makes it different from much work on the 'lived body' (see Chapter 4), is that the focus is on the *practices* which enable the enactment of particular bodies rather than the intentionality and lived experience of the person who might be said to be enacting such practices. Let us explore this study in more detail so that you can get a flavour of how you might use ANT and the concepts associated with it for exploring bodies-in-process.

DOING HYPOGLYCAEMIA

Hypoglycaemia is the name given to the condition, often associated with diabetes, in which a person has an abnormally low level of blood sugar. We might therefore think of hypoglycaemia as a condition of the 'molar' body brought on by the effects of hormones (insulin, for example), or practices which might modify or alter the 'molar' body (such as diet and exercise, for instance). This would be the common understanding of hypoglycaemia, which is embedded within biomedical understandings of diabetes. This is the body as the object and target of particular medical knowledges and practices. Hypoglycaemia is contained or hidden with a singular body and registers as to a blood sugar level below 45 mg/dl. However, rather than simply ask 'what' hypoglycaemia 'is' we might also ask, how it is 'done' performed or enacted. As Mol and Law argue, 'as part of our daily practices, we also do (our) bodies. In practice we enact them' (2004: 45). We bring particular bodies into being and the production or performance of such bodies is intimately connected to the practices, techniques and artefacts which make different bodies possible. In this approach to corporeality or materiality bodies are real but they are also made, remade and even unmade. They are literally brought into being through practices and modes of enactment. They are, most importantly, *made differently* within different practices and body assemblages. This is the focus of ANT and work that moves away from the body as substance to exploring the body-in-process.

Mol and Law suggest that one way of doing hypoglycaemia is by measuring blood glucose levels. This might be enacted through pricking the finger to draw a minute level of blood that can then be subjected to a measuring device which will quantify such levels. Another practice might be what they term 'intro-sensing' (Mol and Law 2004: 48), which is a mode of self-awareness which is oriented towards detecting or sensitively feeling bodily changes. This mode of enactment is literally felt through a sentient body whereby changes might register in a multitude of ways. Mol and Law suggest that the practice or enactment of intro-sensing requires a 'semi-permeable' bodily boundary which is open and extends to others, human and non-human. This might include the surroundings, other people, artefacts, machinery, objects such as food and so forth. This mode of enacting might involve other practices, such as counteracting through eating food, avoiding exercise, or producing the effect of hypoglycaemia through administering insulin. Thus, they argue that hypoglycaemia is done in a variety of different ways: it is measured, felt, countered, avoided and produced. It is not simply *done* within a molar body, but 'incorporates bits and pieces of the world around it, while its action may be shifted out of the body, ex-corporated' (Mol and Law 2004: 53). Hypoglycaemia is, therefore, within this model, not one thing or substance, but rather enacted or brought into being in a variety of different ways which are not necessarily coherent or without tensions. The distinction between what is inside (a molar body) and what is outside is therefore put into question. The body is reformulated as a body that always extends and is augmented by its conjoining with other objects, human and non-human. Each specific coupling or entanglement produces or brings hypoglycaemia into being as a rather distinct, yet different, object. What we witness is *multiplicity*, rather than fixity and the assumption of one static, underlying object: hypoglycaemia.

THE BODY MULTIPLE

> Objects come into being – and disappear – with the practices in which they are manipulated. And since the object of manipulation tends to differ from one practice to another, reality multiplies.

Mol, *The Body Multiple*

The concept of *multiplicity* is central to the work of Anne Marie-Mol, who has written an important book on the body titled *The Body Multiple: Ontology in Medical Practice* (2002). One of the key assumptions of this work is that there is not one singular medical object but rather a range of practices that produce a variety of different objects. The focus in this work is on the *practices* themselves and this shift in focus shows how 'what we might think of as a single object may appear to be more than one' (Mol 2002: vii). She uses the term or concept of the 'body multiple' to refer to the *multiplication* of objects that appear when we focus upon the practices that enact them. The emphasis in this work is on how disease is done in different practices and what the complex relationships are between the different objects that are produced-in-practice. This shifts the focus to a *lived body* (see Chapter 4) that is not simply 'intended' or accorded significance in different narratives and accounting practices (with a focus usually on words and meaning) to exploring how the body is enacted and performed in specific ways. This does not privilege the molar, singular body but instead explores the body as an open system that connects with others, human and non-human. The body is extended to include how it becomes connected up to techniques, artefacts and practices which produce particular kinds of object and entities. There is, therefore, no singular object – hypoglycaemia, for example – that stays within the 'molar' body. The body is always relationally connected, it never just *is*. However, although with a focus upon practice, objects and realities are multiplied, this does not imply that this simply results in fragmentation or pluralism. Mol argues that one of the miracles to be explained is how divergent objects are coordinated so that they 'hang together'. They do not 'hang together' as one coherent object but, rather, are aligned and translated in such a way that contradictions and tensions are made to matter in very particular ways.

Mol focuses upon the example of arteriosclerosis to illustrate how divergent objects produced through heterogeneous practices are coordinated or made to matter in very particular ways. This might include the enactment of arteriosclerosis as pain whilst walking, not being able to climb stairs, or push supermarket trolleys, or walk the dog, as cold feet, weak pulsations, a thickened intima (arteriosclerotic walls), or possibly as a leg amputation, for example. All of these ways of enacting arteriosclerosis are enabled by and require different techniques and artefacts: a microscope, stethoscope, hand with a sensitive touch, a sharp knife, various instruments, recording and measuring devices, machinery, others (human and non-human) and so on. There is not one reality waiting to be discovered but rather a multiplicity of objects that might cohere but might co-exist in contradiction and even discord. The aim of Mol's research which focuses on how arteriosclerosis is enacted is to 'study the multiplication of a single disease *and* the coordination of this multitude into singularity' (2002: 82). The coordination of multiple objects so that they 'hang together' is not due to the discreteness of the object itself, but rather to the strategies and practices which distribute the objects across different sites, locations, activities, experts and interventions. This might accomplish both a 'hanging together' but also a 'keeping apart' of any tensions and contradictions. In this sense 'reality' or 'the body' is distributed so that 'arteriosclerosis enacted in the process of deciding that an operative intervention will be done differs from the arteriosclerosis enacted during the operation' (Mol 2002: 94). As Mol argues, it is the distribution of objects or the body across different sites and practices that 'separate out what might otherwise clash' (2002: 115). She argues that rather than thinking of objects as distortions of the 'true' object, or choosing which object we think arteriosclerosis should be, we should utilize different concepts. These include 'discord, tension, contrast, multiplicity, interdependence, co-existence, distribution, inclusion, enactment, practice, inquiry' (Mol 2002: 180). In this sense practices do not simply describe the body, but rather create what the body might become, and in that sense both enact and have the potential to do the body differently.

SOCIALIZED BIOLOGY

In the concluding part of this chapter we will explore the implications of this work on the body-in-process and the central role of practice and enactment in thinking about the corporeal, somatic and material basis of our bodies. In this sense, we will take the discussion back to what we commonly conceive as our 'molar' bodies, and rethink their materiality in light of the discussions in this chapter. We commonly conceive of the materiality of bodies as something that is fixed, with 'culture' or the 'social' being the aspect of experience that is more fluid and subject to change. This is mirrored by the mind–body dualism that tends to view the body as an inert, machine-like entity with the mind considered to be more subject to cultural or social influence (see Introduction). As we have seen, the molecular gaze of the new biological and life sciences disrupts this view, and, instead, views biological and neurological processes as fluid, plastic and subject to modification and change. This is one way of approaching the somatic or corporeal that challenges the idea that the body is a self-enclosed system and opens and extends the body in such a way that it could be said to be in-process. However, there might be other ways of rethinking materiality in light of the studies reviewed in this chapter that would not simply reduce the body to a synthetic, disembodied medium. As we have seen, this is one of the main critiques of the reduction of the body to information or code (Hayles 1999; Kember 2003). If we focus upon practice and enactment we can start to think about how different practices perform or 'do' the body differently. As John Law (2004) has remarked the focus upon process and practice is on how realities are both made and *re*made. These differences (in realities) are not simply written upon the molar body, but are inscribed into and upon the different practices of everyday life through which bodies are made and remade. In order to illustrate this I want to turn to a study that was carried out with the purpose of thinking about how the phenomenon of voice-hearing is both done (within psychiatric practice, for example) and is being remade and done very differently in practices enacted within the Hearing Voices Network (see Blackman 2001, 2007a). This will lead us to consider the concept of *socialized biology* as a way of thinking about the entanglement and coupling of nature and culture (*naturecultures* – see Haraway 2004) to the extent that we really can no longer hold on to them as pre-existing, separate entities. Another term that we might use to refer to the co-constitution of nature with culture is *enacted materialities* (Law 2004). This term, like 'naturecultures' and 'socialized biology' reflects the fact that the somatic and corporeal are performed and brought into being as much as they are real. They are both real and *made*.

THE BODY AND PSYCHIATRIC CULTURE

The hearing of voices is generally regarded as indicative of mental illness. Indeed, such an interpretation is central to the diagnostic systems of psychiatry and to most psychological forms of treatment. However, there is evidence to suggest that hearing voices is far more common than believed, and that those who develop non-psychiatric explanations of their voices may live with them quite well. This belief forms the basis of an international network of alliances between service users, professionals and families, carers and friends of voice-hearer's known as the *Hearing Voices Network* (HVN). The HVN originated in the Netherlands and is associated with the pioneering work of two Dutch psychiatrists, Marius Romme and Sandra Escher (1993). The beginnings of the network are linked to Marius Romme appearing with one of his patients, Patsy Hague, on a popular Dutch television programme. She talked about her experience and theory as to why she heard voices. The response to the programme was overwhelming, with over 700 voice-hearers contacting Romme. He established that there were many people who heard voices who had never been in touch with psychiatric services. Romme's interest shifted to a focus upon the kinds of non-psychiatric explanations and coping strategies these individuals had developed to manage their voices. He found that people enacted their voices through very different practices and techniques and that these created the possibility of hearing voices in very many different and diverse ways. This focus upon the different practices and techniques that enact voice-hearing in different ways became the subject of a study which considered the implications of this finding for the status of biological explanations of voice-hearing (Blackman 2001, 2007a).

As we saw in the discussion of depression and the role of SSRIs (anti-depressants), biomedical explanations of mental ill-health focus upon the role of biochemical neurotransmitters in the production of particular kinds of symptom. Within this medical model, voice-hearing is often considered a first rank symptom of the discrete disease entity schizophrenia. Although there is controversy over the role of which transmitters and pathways might be involved (dopamine has been the most discussed), it is assumed that voice-hearing is a sign and symptom of psychosis (that a person has lost touch with reality). The biomedical focus is one, therefore, that is based upon a framing of the self and the body that reduces voice-hearing to a neurochemical imbalance in the *brain*. This is a 'molar' view of the body in which the body is reduced to the brain, understood through particular neurochemical explanations. This is what Nikolas Rose terms 'somatic individuality' (2004: 109): a view of the body where it is defined solely in bodily or particular kinds of biological terms. This is also known as an *essentialist* view of the body (see Chapter 1). However, if we shift our focus to *practice*, rather than the body as substance, somewhat different findings emerge. This was the focus of a study which explored some of the different practices that were mobilized by members of the HVN and how these practices produced the voices as rather different kinds of entities (Blackman 2001).

Many members of the HVN have been failed by psychiatry. Although they may have taken, or indeed be taking, anti-psychotic medication, often the voices have remained. The HVN have provided different sites, locations and practices through which voices can be enacted differently, thus providing relief from distressing and often abusive voices, or enabling the voice-hearer to have more choice and control over when and where they hear the voices, and even which voices they listen to. What is remarkable about the HVN is that it provides a network or platform through which the voices themselves potentially transform. What might have started as abusive voices may become less abusive, or disappear altogether, perhaps to be replaced by voices that the voice-hearer may wish to keep as they function much like good friends and as a form of social support. These changed experiences of the voices themselves are enacted through transforming the person's relationship to their voices. Within psychiatric practices the voices are viewed as symptoms of disease and illness which should primarily be ignored, not focused upon and acted upon through the administration of anti-psychotic medication. The HVN have reversed this relationship, instead encouraging the voice-hearer to listen to their voices, recount them (to others), write them down and think about how their content might relate to their life-history and circumstances. The HVN encourage voice-hearers to develop a 'frame of reference'. These frameworks incorporate different ways of understanding and enacting one's identity as a voice-hearer.

A popular frame of reference or practice for enacting the voices is one that views the voices as a gift or sensitivity rather than a sign of illness or disease (in the brain). These are spiritual sets of practices which predominantly view the voices as a sign of telepathy or mediumship (that the person is acting as a conduit for voices that are being passed or communicated by others – usually the dead). Although these explanations might seem far-fetched what is remarkable, regardless of whether you believe in the possibility of telepathy, is that the practices make or do the voices as very different kinds of entity. If we suspend our belief in the truth-status of such practices, and rather focus on what they do, we find that the practices enable the embodiment of the voices as rather different kinds of phenomena. Rather than being feared, causing distress and enacted as signs of illness and disease, they are embraced, focused upon, accepted and viewed as offering the potential for psychic reverie. The voices become the object of practices which attempt to achieve calm through meditative and visualization techniques – what I have termed practices of psychic isolation, techniques of symbolism and an economy of physical and mental regimens (Blackman 2001). The voices are literally enacted and embodied as very different kinds of object, and these differences in voice-hearing are inscribed and written upon the practices themselves. So, for example, voice-hearers have described their changed experiences of their voices, with the reported changes possibly including the content, where they are heard, changed feelings of control and feelings of calm rather than dread, terror, shame, confusion and anxiety, for example. This suggests that by focusing upon *practice* and *enactment* we can start to see the complex relationality of the body, affect and culture. Rather than the idea of the biological and cultural as separate entities that somehow interact, we move to a notion of entangled processes that meet, enfold, invigorate and pass with the effect that they do not remain the same.

Conclusion: Enacted Materialities

The work reviewed in this chapter suggests that the relationship between the body and culture is very complex, and that in order to understand this complexity we need new tools for analysis and new concepts and vocabularies for 'thinking through the body'. We have focused in this chapter on the importance of *process* and *practice* for inventing just such vocabularies. The body is never 'simply biological' or socially constructed. These paradigms are both reductive, as we have seen, and are part of the problem for body theory and the directions it is currently taking across the humanities. We will engage with some of these directions in the Conclusion. We have seen in this chapter how although molecular biology is providing a shift in how the body is understood (not as a closed system, for example), it tends to reduce the body to *code* or *information*. The focus on *enactment* and *practice* brings with it a renewed opportunity to view the body as open to being affected and affecting, but not simply to view biology as code. We have seen in work on the body from the

130 the body: the key concepts

perspective of enactment that what we might term 'biological experiences', such as hypoglycaemia, or voice-hearing, are not fixed and stable kinds or entities; they shift and change in the different practices that perform them or bring them into being as particular kinds of object. Biological processes pass through practices meaning that they do not remain the same. Biology is socialized or enacted: it is both *real* and *made* and requires a more complex relational approach to understand its entanglement. It is not just that the brain manifests a plasticity or fluidity, but that what we understand the biological to be is always subject to change. Biology is *socialized*. This is what Mariam Fraser terms the 'generative force of matter' (2001: 621) where she develops Vicky Kirby's (1997) concept of the 'mediated nature of nature' (quoted in Fraser 2001: 618). This does not mean that culture and the body are the same thing so that they can be collapsed into one another. However, it does mean that the separation of culture and the body is problematic and asking the question 'Is this culture or the body?' has excluded and silenced a whole array of questions that open the body up to new kinds of analysis and questioning.

CONCLUSION: IMAGINING THE FUTURE OF THE BODY WITHIN THE ACADEMY

INTRODUCTION

The emergence of the body as an important focus and object of study within the humanities has begun the important work of reformulating many concepts that have been integral to contemporary theories. These include power, subjectivity, agency, technology, the human, life, the social, biology and more besides. Although I do not want to chart a linear trajectory underpinning such a journey, one common theme that runs throughout this book is a concern with exploring what bodies could, might and indeed arguably have become. This is particularly so if we resist the temptation to think of them as entities which are singular, bounded, molar and discretely human in origin. This naturalistic view of the body has dominated the splits between the natural and human sciences, and is one which is being refused and rethought in work across the humanities that take body matters seriously. 'Thinking through the body' inevitably opens academic reflection on to areas of study which at times might seem bizarre, perplexing and of tangential concern to theorizing the social. However, what we also find is the very notion that there is a homogenous social domain from which we can add to concepts and theories from the life and biological sciences is somewhat misguided. This has been one guiding premise from early work in feminism and the sociology of the body which has refused the assumption that the psychological, biological and social are discrete entities that somehow interact. The problem of the positing of an 'interaction effect' (Riley 1983: see Chapters 1 and 2) to explain human subjectivity has been refused, although the question of exactly how we can theorize processes that are thoroughly entangled and interdependent still presents an important challenge to contemporary theory.

The chapters in this book in different ways explore concepts that attempt to do just this, and importantly reinvent the questions we might ask about bodies. The certainty of method is abandoned or silently put to one side, alongside the recognition that asking exactly what bodies are or what a body is no longer seems tenable. Different questions and modes of inquiry are called for. The question of what a body is presumes the body as substance, and that this realm (what we might term the corporeal, somatic or material) can be known and contained through the adoption of particular kinds of method. However, what we see throughout the book are attempts to develop methods that are not based on certainty, and which at times are attempting to render visible aspects of experience which might usually remain silent, unnoticed and in the background. Ann Game (2001) reinvents concepts such as *attunement* from spiritual practices that include Buddhism to describe the 'sensitive feel' that was required in her relationship with her paralysed horse and its eventual recovery. These aspects of experience suggest something else about bodies: not materiality, but perhaps an *immateriality* that is felt and registered but cannot easily be seen, known or understood. Early psychological experiments, such as the case of Hans the horse, testify to the devices, which might unwittingly make these immaterial processes visible. Hans's astounding abilities to solve multiplication puzzles were interesting because they threatened the notion of the 'knowing subject'. The experimenter was unknowing in that he did not know he was performing minimal body movements, and these movements could not be controlled within the experimental situation in which Hans found himself. They were the unexplained background, what Latour (2005b: 230) calls 'the everything else', that was present and that could not be eliminated. The re-engagement with anomalies such as Hans the horse by body theorists has moved the terrain of body studies in new directions (Despret 2004a and b). These new directions are the subject of this concluding chapter, which will give a flavour of what are emerging as important concerns for the future of body theory.

THE AFFECTIVE BODY

The use of the term 'body theory', however, is to be approached with caution. We do not start with bodies as a key focus, but, as we have seen throughout the book, concerns about lived experience, sleep, marching, dance, identity, eating disorders, technologies, the placebo effect, communication, body language, performance, emotion, twinning and cloning, the senses, the mouth and health and illness inevitably involve talk of the body. But the body that organizes such diverse practices and areas of experience is a body that is open, relational, human and non-human, material and immaterial, multiple, sentient and processual. The body is not a thing to retreat to, a material basis to explain how social processes take hold. The body is *in process* and is assembled and made up from the diverse relays, connections and relationships between artefacts, technologies, practices and matter which temporarily form it as a particular kind of object. However, even the term 'it' implies a form or shape that can be easily recognizable as a body. What is clear from the book thus far is that talk of the body extends to talk of body assemblages that might not resemble the molar body in any shape or form. The body has been extended to include *species bodies, psychic bodies, machinic bodies, vitalist bodies* and *other-worldly* bodies, which do not conform to our expectations of clearly defined boundaries between the psychological, social, biological, ideological, economic and technical, for example. Bodies are processes that are articulated and articulate through their connections with others, human and non-human. In this sense, if there is one guiding principle towards which work on the body has moved it is the assumption that what defines bodies is their capacity to affect and be affected. The focus upon the affective capacities of bodies, human and non-human, is extending the terrain of body studies in new and exciting directions. Although it is arguable whether such a focus will achieve the paradigm shift associated with the turn to language and the subsequent turn to the body within the humanities, some are proclaiming 'the turn to affect' as extending some of the trends we find throughout the book, directly and indirectly, in innovative ways (Hardt 2007).

Patricia Clough (2007: 2) relates the human body's capacity to affect and be affected to part of our 'self feeling of being alive'. This feeling of vitality could usefully be connected to work on the somatically felt body that we have explored throughout the book, and which is usually dismissed as being about a realm of feeling that is largely automatic and involuntary (see Chapter 2). These are aspects of bodily affective capacities which are being made central to contemporary theorizing rather than relegated to a mechanistic, largely physiological body. These studies arguably reveal aspects of our embodied experience that have been submerged and forgotten. The problem of the 'inert body' was recognized as part of the problem that the sociology of the body had to contend with from its inception. As Lash argues, 'the body should possess some positive, libidinal driving force' (1991: 277). This is now a central concern and is framing research directions that are producing concepts, lexicons and vocabularies for framing feeling and affectivity as important objects of study (Fraser *et al.* 2005; Blackman and Cromby 2007; Blackman 2007c). Hardt (2007) suggests that the so-called affective turn extends work that focuses upon the body whilst introducing an important shift. This shift is focused upon how we 'think' the relational dimensions of corporeality; (of what a body can do, for example), without sidelining the role of *power* and *regulation*. We saw in Chapter 1 how the regulated and regulating body was an important focus of work that emerged

within the sociology of the body, and which particularly developed the work of
Michel Foucault and the *disciplined* or *docile body* (see Chapter 1). This work was
seen to be useful as it presents the body as malleable, as an unfinished entity that can
be sculpted, moulded, altered and transformed through disciplinary practices. Power
is seen to work through inculcation (requiring the participation of subjects) rather
than imposition (which assumes power as top-down and repressive). However, we
explored in the same chapter how Foucault's studies arguably ignored an important
aspect of corporeal dimensions of power – that is, that the body is not simply inert
mass but has vitality. We explored this in relation to military drill, which makes
visible aspects of corporeality (or immateriality) that are missed by Foucault's studies.
The concept of *muscular bonding* (McNeill 1995) points towards the somatically felt
dimensions of rhythm and keeping in time which literally make people feel good
and propel them to potentially invest in particular practices. This introduces *energy*
and *creative motion* into conceptions of materiality and immateriality and moves
studies of corporeality into areas that are difficult to see and know.

IMMATERIAL BODIES

Many contemporary authors are arguing that *immateriality* is a defining theme
that organizes a variety of different practices in advanced liberal societies. These
include biotechnological practices that reduce the body to code and information
(see Chapter 5) and work practices that increasingly require forms of affective labour
– 'affective labour' being a term used to describe the skills and emotional or affective
capacities that are increasingly required within informational economies, such as the
'service' sector (Hochschild 1983), the culture industries (particularly advertizing
and marketing), healthcare, welfare, fashion modelling (Wissinger 2007) and so
forth. Affect management, for example has entered the workplace and is recognized as an important process through which specific entities, such as groups, teams
and managers, can be brought into being as particular kinds of worker (Barsade
and Gibson 2007). 'Affective labour' is a term that aligns the central importance
in these economies of communication, knowledge, information and affect (Hardt
and Negri 2000; Thrift 2004, 2007; Lazzarato 2004). Examples might include
fashions, tastes, public opinion and consumer norms (cf. Clough 2007; Lazzarato,
2004). These processes are seen to be largely immaterial: that is, they rely upon the
circulation of information, ideas, images, and affect through our bodies in ways that
are difficult to see and measure, particularly in terms of their value or contribution
to labour economies (Clough 2007). One example of this might be the affective
skills and capacities that allow workers to organize and be organized as particular

kinds of flexible, mobile, multiple yet self-regulating subjects (see Blackman 2005). The ability to embody such worker subjectivities is often presented as down to the internal psychological capacities or resources that the subject is required to develop (such as emotional intelligence), reliant upon an understanding of the body as a closed system. This presumption obscures the ways in which workers' affective capacities are modulated and transformed through their conjoining with particular kinds of practices and techniques that extend the body beyond itself.

MODULATION

The term 'modulation' was developed by Deleuze (1990) to refer to the new modes of power that are central to the workings of what he termed 'control societies': those societies in which power no longer primarily operates through institutional practices and techniques inscribed within spatial enclosures, such as the prison system, the factory, the school etc (see Chapter 1 for a discussion on disciplinary power). Modes of modulation continually change from moment to moment and cannot be captured within environmental enclosures or sites. In this sense, power operating through modulation works through practices that can never simply be thought of as rational, cognitive, or ideological (6 2007). Modulation as a mode of power works through affect, emotionality, contagion and intensity, perhaps revealing the 'market-driven circulation of affect and attention' (Clough 2007: 19). Modulation is thus more relevant to studies of new forms of production and consumption that circulate within *informational, knowledge* or *networked* societies. Modulation refers to the horizon of practices, techniques, artefacts and objects which form, vary, alter and shape matter (understood as information) through operating upon an intensive and affective realm. Modulation organizes, orchestrates, mobilizes and amplifies this affective realm or register. The key concept is *movement* – of practices, bodies and matter. As Clough argues, 'The target of control is not the production of subjects whose behaviours express internalized social norms; rather, control aims at a never-ending modulation of moods, capacities, affects and potentialities, assembled in genetic codes, identification numbers, ratings profiles, and preference listings, that is to say, in bodies of data and information (including the body as information and data)' (2007: 19).

Within these conceptions, as we have already seen, the body is considered open and relational (affecting and affected), rather than closed and static. Bodies are viewed as always in a process of *becoming*. Matter is thus considered dynamic rather than fixed and closed (particularly in this example at the molecular level), and modulation refers to the processes through which matter varies and changes. As Clough argues in relation to these new configurations of 'bodies, technology and matter', we need tools and concepts which move beyond an economy of production and consumption to explore how 'bodily capacities are modulated and augmented through their conjoining with technologies' (2007: 2).

This is an important direction in work on bodies that signals an engagement with theories and concepts from the psychological, biological and life sciences that were once seen as adversaries within the humanities. As we saw in Chapter 1, work on bodies that originated within social constructionist, or cultural inscription paradigms tended to dismiss the biological and psychological sciences as reductionist and essentialist. One trend of recent work on bodies rejects the claims of such perspectives and argues instead that an engagement with such knowledge practices is important in understanding how bodily and affective capacities are altered, transformed and augmented through their cooption within different practices and body assemblages. We have already considered some examples of work on bodies that takes the psychological sciences as an important site for reformulating human subjectivity (see Chapter 2). These include Elizabeth Wilson's work on neurology (2004), Teresa Brennan's work on 'emotional contagion' (Brennan 2004) and Vicianne Despret's work on emotion (2004a and b). All of these studies, in one way or another, conclude that bodies are 'psychologically attuned' in that they are always open to being affected and affecting and that these processes are inscribed at the level of hormones, neurotransmitters and the nervous system, for example. Rather than being viewed as fixed and internal, psychological matter is seen as dynamic, mutable and both inside and outside. One example, pertinent to this discussion is the concept of *biomediation* that has been developed by an Australian cultural theorist, Anna Gibbs (2002; Angel and Gibbs 2006).

BIOMEDIATION

The focus of this work is on how affect is transmitted by media technologies and practices through a form of *contagious communication*, particularly on how the *face* operates as a key device through which affect is transmitted and thus provides an interface through which modulation occurs. Angel and Gibbs develop the work of Silvan Tomkins (1962; Sedgwick 2003), the American psychologist, who argued that 'the face is the primary site of affective communication' (2006: 25). They trace this interest in the face back to the writings of the evolutionary biologist Charles Darwin (1859) who, as we saw in Chapter 1, is more associated with the *naturalistic body*. Darwin (1872) was interested in how different facial expressions associated with a range of different emotions, such as fear, anger, surprise and disgust, were underpinned by particular physiological responses by the autonomic nervous system. These might include sweating, muscle tremors, increased digestion and respiration etc. This work was taken up by Tomkins who argued that there are 'nine discrete innate human affects' (2006: 26) and that these can be amplified, magnified and modified through communication processes.

The term 'biomediation' is associated with the writings of Eugene Thacker (2003), who has argued that what we see through the coupling of technology with bodies is the shaping and conditioning of 'biology' in which media practices reconfigure the human body. This reconfiguring primarily occurs through an affective realm with the result that our capacity to be affected is coopted and modulated so that we are inserted or 'plugged-in' (Latour 2005a) to particular media flows. The face is the visible representation of the complex association of elements that make up particular media flows thus becoming 'a technology through which these processes are communicated and enacted on the social body' (Angel and Gibbs 2006: 27). Different faces (the face of David Beckham, Winston Churchill, George Bush, John Howard and Nelson Mandela, for example) condense assemblages of different visual images, sound bites, ideas, practices, texts and objects that are transmitted with an immediacy through their iconic images. The face (and particularly the use of the close-up shot) is therefore considered a 'major interface for the transmission of affect which binds human beings in relationship with each other' (Angel and Gibbs 2006: 29). As we can see, this is not simply the transmission of emotional expression but the transmission of assemblages of objects, practices, ideas and beliefs through the amplification and modulation of affective capacities. These are capacities that are felt and sensed but which are difficult to articulate and understand.

CORPOREAL THINKING

This focus on 'contagious communication' develops in many ways work that we explored in Chapter 2 on 'emotional contagion'; that is, the idea that emotions can be passed between people and do not inhere within the private, individualized self. This moves well beyond the assumption that language or discourse constructs the body (see Chapters 1 and 3) towards the notion that bodies are made and remade through the mediation and modulation of biological capacities that are always dynamic and in relationship with what we might term 'the outside': machines, practices, technologies and so forth. In this sense, 'biology', or matter, is not an entity but is defined as a relational, dynamic process which is enfolded with the 'outside'. The use of the term 'fold' points towards the complex entanglement and interweaving of the inside with the outside to the extent that it is impossible to make such distinctions or differentiations (see Rose 1996). This focus on what we might term an 'affective register' (Thrift 2004) is therefore a key focus of much work that is taking the study of the body and bodies in new and exciting directions. This work suggests that language is not the key mode of communication, and that much of what passes as communication inheres within a realm that is difficult to see, understand and articulate (see Chapter 2). This work also suggests that we are far from unitary subjects but are multiple and defined by our capacity to be *affected* and *affect* others. It is also often through processes of non-conscious perception rather than through rational and conscious deliberation that the registering of this multiplicity is primarily felt and sensed (Thrift 2004; Connolly 2002). This is a form of 'corporeal thinking' (Thrift 2004: 65) rather than thinking separated from an inert physiological body (mind–body dualism).

Conclusion

Although this new trend of body theory focuses upon the importance of considering affective energies and creative motion, the body as process and in-movement, we would still be wise to retain the importance of attending to the concept of *subjectivity*, which refers to the ways in which individual subjects attempt to 'hang together' a coherent sense of self in the face of multiplicity (see Henriques, Venn, Urwin and Walkerdine 1998). The focus on subjectivity in body theory was considered in Chapters 3 and 4, which took the individualized self as something to be achieved or accomplished through particular strategies, techniques and devices. This focus on subjectivity perhaps retains the importance of what we might term 'psychic bodies': the psychic defences and strategies that subjects develop in order to survive and live in the world. Perhaps rather than thinking about the subject as multiple we also need to think about how singularity is *lived* in the face of multiplicity (see Walkerdine 2007; Lee and Brown 2002). In this sense, the psychic or psychological cannot simply be replaced by 'biology' understood as dynamic matter, unless we are also prepared to consider bodies as being psychologically attuned in ways that cannot be reduced to the neurophysiological, endocrinological or neurological. This is also an important direction for body theory that does not assume the body as substance or body as organism but does point towards the work that is required to assume a sense of coherence in the face of *process, movement, multiplicity* and *becoming*. As Mol and Law (2004: 57) suggest, 'keeping ourselves together is one of the tasks of life', and this requires new 'conceptualizations of what it might mean to hold together' (Law and Mol 2002: 10). How we can be 'one yet many' and 'multiple yet singular' is a key question for body theory and one that promises to bring the study of brain, body and culture into new, perhaps as yet unforeseen, areas and alliances.

QUESTIONS FOR ESSAYS AND CLASSROOM DISCUSSION

INTRODUCTION: THINKING THROUGH THE BODY

1. Discuss the differences between approaching the body as a substance or entity, and the approaches which focus on 'the body' as a process.
2. What is meant by the concept of dualism?
3. Describe some experiences of 'mixing' or attunement that you may have experienced in your own lives. These experiences might include those with both human and non-human Others.
4. Discuss some of the problems with the maxim 'mind over matter'.
5. What are some of the problems with the concept of the 'natural body'?

I REGULATED AND REGULATING BODIES

1. To what extent has the body been an 'absent presence' within sociological theorizing?
2. What does it mean to study bodies as 'unfinished entities'?
3. What are some of the problems with models of cultural inscription?
4. Discuss the differences between a naturalistic body paradigm and a socially constructed body paradigm.
5. To what extent does Williams's (2005) more embodied approach to the 'sleeping body' overcome the problems presented by these two paradigms?

2 COMMUNICATING BODIES

1. Discuss the differences between models of 'social influence' and more embodied approaches to communication, particularly in the context of human and non-human relationships.
2. How does Despret's (2004a and b) conception of becoming transform how we might understand non-verbal communication?
3. Why do you think a discourse of authenticity is so prominent within popular and consumer culture? Illustrate your answer with some examples.
4. To what extent does the phenomenon of emotional contagion threaten the concept of affective self-containment (Brennan 2004) and what are the implications of this work for body studies?
5. Describe some examples of 'the somatically felt', networked or vitalist body.

3 BODIES AND DIFFERENCE

1. Why does Skeggs (1997, 2004) argue that 'respectability' has become a key marker of social distinction in advanced liberal democracies?
2. To what extent can class be considered a form of corporeal capital?
3. Discuss some of the different ways in which a concept of 'bodily affectivity' has been used to understand the enactment of difference, and what does this introduce to studies of bodily dispositions?
4. Discuss some of the different ways in which the concept of feminine becoming has been taken up by different scholars to interrogate the politics of female bodies.
5. What are some of the problems and possibilities with the concept of gender performativity?

4 LIVED BODIES

1. Why is history important for understanding sense organization?
2. What does the concept of skin knowledge contribute to studies of the senses?
3. What does the concept of abjection reveal about the importance of bodily borders and boundaries?
4. What is meant by the 'articulated body'? Illustrate your answer with examples.
5. What can studies of 'self-health' or 'somatic individuality' reveal about the body as it is lived in health and illness?

5 THE BODY AS ENACTMENT

1. What do studies of 'bodies-in-process' reveal about corporeal matters?
2. What is useful to body theory in thinking about bodies as being defined by their capacity to affect and be affected?
3. Critically examine the concept of 'bodies without organs'. What are some of the problems and possibilities with this concept? Give some examples to illustrate your answer.
4. What does it mean to enact bodies? How does the concept of enactment blur the distinction between what is inside and outside, the human and non-human and the singular and multiple?
5. Discuss the usefulness of the concept of 'socialized biology' for our understandings of the lived body.

CONCLUSION: IMAGING THE FUTURE OF THE BODY WITHIN THE ACADEMY

1. To what extent is the concept of the affective body reformulating our understandings of embodiment and disembodiment?
2. To what extent is the concept of modulation useful for understanding the relationships between bodies and power?
3. How does the concept of biomediation challenge the historical separation between the humanities and the life and biological sciences?
4. To what extent are understandings of subjectivity still centrally important for body studies?
5. What do you think are the important directions in which body studies should go in the future?

ANNOTATED GUIDE FOR FURTHER READING

Braidotti, R. (2002), *Metamorphoses: Towards a Materialist Theory of Becoming*. Oxford: Polity Press.

This is a hugely influential book that paved the direction for 'corporeal feminism', which has attempted to identify the importance of Deleuzian concepts for reformulating bodily matters in the context of the politics of female bodies.

Butler, J (1990), *Gender Trouble: Feminism and the Subversion of Identity*. London and New York: Routledge.

Butler, J. (1993), *Bodies that Matter: On the Discursive Limits of 'Sex'*.

The American academic Judith Butler, a philosopher by training, developed an understanding of gender performance and performativity for interrogating how gender is enacted through a process of normalization. Her work has been hugely influential across the humanities and brings together Foucauldian work on discourse with psychoanalytic approaches to subjectivity.

Burkitt, I. (1999), *Bodies of Thought: Embodiment, Identity and Modernity*. London: Sage.

This important book, written by a British sociologist, examines embodiment by drawing primarily on phenomenological traditions and particularly the scholarship of the French philosopher Merleau-Ponty.

Clough, P. (ed.) with Halley, J. (2007), *The Affective Turn: Theorizing the Social*. Durham, NC and London: Duke University Press.

This book outlines the importance of affect and the affective body for reconfiguring how we might understand important sociological concepts such as power, subjectivity, technique and practice.

Crossley, N. (2001), *The Social Body: Habit, Identity and Desire*. London: Sage.

This book, and Crossley's subsequent development of approaches to the body within sociology, develops work on body techniques for understanding how bodies come to be lived in such a way that they appear natural and 'un-thought'. Crossley primarily draws on the work of the French sociologist Pierre Bourdieu and phenomenological traditions which focus on the importance of feeling, affect and desire.

Csordas, T. (ed.) (1994), *Embodiment and Experience: The Existential Ground of Culture and Self*. Cambridge: Cambridge University Press.

This book, edited by the important American anthropologist Thomas Csordas, brings together a collection of chapters by scholars attempting to think the 'lived body' beyond either biological or social/discourse determinism.

Deleuze, G. and Guattari, F. (1987), *A Thousand Plateaus: Capitalism and Schizophrenia.* Minneapolis: University of Minnesota Press.

This book is where Deleuze and Guattari most convincingly develop their concept of 'BwO' (Bodies without Organs).

Despret, V. (2004), *Our Emotional Make-Up: Ethnopsychology and Selfhood.* New York: Other Press.

The French philosopher Vinciane Despret outlines her innovative theory of becoming in the context of refiguring emotion and the body. Her work is very influenced by Actor Network Theory (Latour, 2005a) and philosophers such as William James (1902).

Elias, N. (1994), *The Civilizing Process: The History of Manners and State Formation and Civilization.* Oxford: Blackwell.

Elias, N. (2000), (2nd edition) *The Civilizing Process: Sociogenetic and Psychogenetic Investigations.* Oxford: Blackwell.

These volumes present a very important historical contribution to the conditions which led to the emergence of the concept of the separate, individualized self.

Foucault, M. (1973), *The Birth of the Clinic.* London: Tavistock.

Foucault, M. (1977), *Discipline and Punish.* London: Allen Lane.

The work of the French philosopher and historian has been hugely important in conceptions of the relationships between bodies and power and the subject and subjectivity. The following books present secondary introductions to Foucauldian studies which are accessible and clearly presented:

Danaher, D., Schirato, T. and Webb, J. (2000), *Understanding Foucault.* London: Sage.

O'Farrell, C. (2005), *Michel Foucault.* London: Sage.

Halberstam, J. (1998), *Female Masculinity.* London and Durham, NC: Duke University Press.

Halberstam, J. (2005), *In a Queer Time and Place: Transgender Bodies, Subcultural Lives.* New York: New York University Press.

Judith Halberstam presents approaches to queer performativity which disrupt gendered binarisms and offer a formulation of 'queer time' for understanding the invention of new models of (queer) kinship.

Haraway, D. (1991), *Simians, Cyborgs and Women: The Reinvention of Nature.* London and New York: Routledge.

Haraway, D. (1996), *Modest_Witness@Second_Millennium.FemaleMan©_Meets_ Oncomouse™.* London and New York: Routledge.

Haraway, D. (2003), *The Companion Species Manifesto: Dogs, People and Significant Otherness.* Indiana: Indiana University Press.

Haraway, D. (2004), *The Haraway Reader.* London and New York: Routledge.

The writings of the American feminist science studies writer (a biologist by training) has been hugely influential in offering new figurations for rethinking bodies beyond singularity and separation (particularly of human/animal and human/machine captured in her figurations of companion species and the cyborg respectively).

Laporte, D. (2000), *History of Shit*. Cambridge, MA and London: The MIT Press.
A very entertaining and thought-provoking account of how the domestication of shit is linked to the historical emergence of the individuated separate self.

Latour, B. (2005), *Reassembling the Social: An Introduction to Actor Network Theory*. Oxford and New York: Oxford University Press.
This is an important book which outlines a method which refuses to accept what it means to offer social dimensions to analyses of bodily matters.

Massumi, B. (2002), *Parables for the Virtual: Movements, Affect, Sensation*. Durham, NC and London: Duke University Press.
This is an important book which has placed studies of affect, sensation and vitalism firmly on the agenda for studies of corporeality within the humanities.

Mol, A. (2002), *The Body Multiple: Ontology in Medical Practice*. London and New York: Duke University Press.
This book, written by the Dutch anthropologist AnneMarie Mol, sets out how a 'praxiographic' approach to the enactment of bodies within medical settings can change how we might think about disease 'entities' such as arteriosclerosis.

Rose, N. (1989), *Governing the Soul: The Shaping of the Private Self*. London: Routledge.

Rose, N. (1996), *Inventing Ourselves: Psychology, Power and Personhood*. Cambridge: Cambridge University Press.

Rose, N. (1999), *Powers of Freedom: Reframing Political Thought*. Cambridge: Cambridge University Press.

Rose, N (2007), *The Politics of Life Itself: Biomedicine, Power and Subjectivity in the 21st Century*. Princeton: Princeton University Press.
Nikolas Rose is a leading British sociologists who has developed the work of Michel Foucault to understand the emergence of the psychological and in his later work the life sciences for the formation of embodied subjectivities.

Shilling, C. (2003), *The Body and Social Theory*. 2nd edn. London, Thousand Oaks, CA and New Delhi: Sage.
Chris Shilling is one of the leading proponents of body theory within sociology and has made important contributions to the sociology of the body.

Thrift, N. (2007), *Non-Representational Theory: Space, Politics, Affect*. London and New York: Routledge.
Nigel Thrift is a geographer who has written an important book bringing together his work on affect for rethinking conceptions of space, time and the body.

Turner, B. (1996), *The Body and Society: Explorations in Social Theory*. 2nd edn. London and Thousand Oaks, CA: Sage.
Bryan Turner is also a leading scholar in the sociology of the body who has written extensively on corporeal matters, particularly in the context of identity and difference.

Williams, S. J. and Bendelow, G. (1998), *The Lived Body: Sociological Themes, Embodied Issues*. London and New York: Routledge.
This is an important book which develops a conception of the 'lived body' for reformulating corporeal matters.

BIBLIOGRAPHY

Aalten, A. (2007), 'Listening to the Dancer's Body' in C. Shilling (ed.), *Embodying Sociology: Retrospect, Progress and Prospects*. Oxford: Blackwell Publishing.

Ahmed, S. (1998), *Differences that Matter: Feminist Theory and Postmodernism*. Cambridge: Cambridge University Press.

Ahmed, S. (2000), *Strange Encounters: Embodied Others in Post-coloniality*. London and New York: Routledge.

Ahmed, S. (2004), *The Cultural Politics of Emotion*. Edinburgh: Edinburgh University Press.

Ahmed, S. (2006), *Queer Phenomenology: Orientations, Objects, Others*. London and New York: Duke University Press.

Ali, S. (2003), *'Mixed Race', Post-race: Gender, New Ethnicities and Cultural Practices*. London and New York: Berg.

Allport, G. (1955), *Becoming: Basic Considerations for a Psychology of Personality*. New Haven, CT: Yale University Press.

Angel, M. and Gibbs, A. (2006), 'Media, Affect and the Face: Biomediation and the Political Scene'. *Southern Review*, 38(2): 24–39.

Anzieu, D. (1989), *The Skin Ego*. New Haven, CT: Yale University Press.

Balsamo, A. (1996), *Technologies of the Gendered Body: Reading Cyborg Women*. Durham, NC: Duke University Press.

Barsade, S. G and Gibson, D. E. (2007), 'Why does Affect Matter in Organisations'. *The Academy of Management Perspectives*, 21(1): 36–59.

Bartky, S. L. (1997), 'Foucault, Feminism and the Modernization of Patriachal Power' in K. Conboy, N. Medina and S. Stanbury (eds), *Writing on the Body: Female Embodiment and Feminist Theory*. New York: Columbia University Press.

Beck, U. and E. Beck-Gernsheim (2000), *Individualization*. London: Sage.

Best, S. and Kellner, D. (2004), 'Biotechnology, Ethics and the Politics of Cloning' in Nico Stehr (ed.), *Biotechnology: Between, Commerce and Civil Society*. New York: Transaction Press.

Bhabha, H. (1994), *The Location of Culture*. London: Routledge.

Blackman, L. (2001), *Hearing Voices: Embodiment and Experience*. London and New York: Free Association Books.

Blackman, L. (2005), 'The Dialogic Self, Flexibility and the Cultural Production of Psychopathology'. *Theory and Psychology*, 15(2): 183–206.

Blackman, L. (2007a), 'Psychiatric Culture and Bodies of Resistance'. *Body and Society*, 13(2): 1–24.

Blackman, L. (2007b), 'Reinventing Psychological Matters: The Importance of the Suggestive Realm of Tarde's Ontology', *Economy and Society*, 36(4): 574–96.

Blackman, L. (2007c), 'Feeling F.I.N.E: Social Psychology, Suggestion and the Problem of Social Influence', *International Journal of Critical Psychology*, 21: 23–49.

Blackman, L. (2008a), 'Affect, Relationality and the 'Problem of Personality', *Theory, Culture and Society*, 25(1): 27–51.

Blackman, L. (2008b), 'Is Happiness Contagious?' *New Formations*. Special Issue on Happiness, guest ed. S. Ahmed, 63: 15–22.

Blackman, L. and Cromby, J. (eds) (2007), 'Editorial: Affect and Feeling'. Special Issue of the *International Journal of Critical Psychology*, 21: 5–22.

Blackman, L. and Walkerdine, V. (2001), *Mass Hysteria: Critical Psychology and Media Studies*. Basingstoke and New York: Palgrave Macmillan.

Bloor, M. and McKintosh, J. (1990), 'Surveillance and Concealment: A Comparison of Techniques of Client Resistance in Therapeutic Communities and Health Visiting' in S. Cunningham-Burley and N. McKeganey (eds), *Readings in Medical Sociology*. London: Routledge.

Bordo, S. (1993), *Unbearable Weight: Feminism, Western Culture and the Body*. Berkeley: University of California Press.

Bordo, S. (1997), 'The Body and the Reproduction of Femininity' in K. Conboy, N. Medina and S. Stanbury (eds), *Writing on the Body: Female Embodiment and Feminist Theory*. New York: Columbia University Press.

Bourdieu, P (1984), *Distinctions: A Social Critique of the Judgement of Taste*. Cambridge, MA: Harvard University Press.

Bourke, J. (2005), *Fear: A Cultural History*. London and New York: Virago Press.

Braidotti, R. (1996), *Between Monsters, Goddesses and Cyborgs: Feminist Confrontations with Science, Medicine and Cyberspace*. London: Zed Books.

Braidotti, R. (1997), 'Mothers, Monsters and Machines' in K. Conboy, N. Medina and S. Stanbury (eds), *Writing on the Body: Female Embodiment and Feminist Theory*. New York: Columbia University Press.

Braidotti, R. (2002), *Metamorphoses: Towards a Materialist Theory of Becoming*. Oxford: Polity Press.

Brennan, T. (2004), *The Transmission of Affect*. Ithaca, NY: Cornell University Press.

Brook, B. (1999), *Feminist Perspectives on the Body*. London and New York: Longman.

Brown, M. (2004), 'Taking Care of Business: Self-Help and Sleep Medicine in American Corporate Culture'. *Journal of Medical Humanities*, 25(3): 173–87.

Burkitt, I. (1999), *Bodies of Thought: Embodiment, Identity and Modernity*. London: Sage.

Butler, J. (1990), *Gender Trouble: Feminism and the Subversion of Identity*. London and New York: Routledge.

Butler, J. (1993), *Bodies that Matter: On the Discursive Limits of 'Sex'*. London and New York: Routledge.

Butler, J. (2004), *Precarious Life: The Powers of Mourning and Violence*. London: Verso.

Butler, J. (2005), *Undoing Gender*. London and New York: Routledge.

Cantril, H. (1940), *Invasion From Mars: A Study in the Psychology of Panic*. Princeton: Princeton University Press.

Charlesworth, S. (2000), *A Phenomenology of Working Class Experience*. Cambridge: Cambridge University Press.

Classen, C. (2005), 'Tactile Therapies' in C. Classen (ed.), *The Book of Touch*. Oxford: Berg.

Clough, P. (2007), 'Introduction' in P. Clough (ed.) with J. Halley, *The Affective Turn: Theorizing the Social*. Durham, NC and London: Duke University Press.

Clough, P. (ed.) with Halley, J. (2007), *The Affective Turn: Theorizing the Social*. Durham, NC and London: Duke University Press.

Collett, P. (2003), *The Book of Tells: How to Read People's Minds from their Actions*. London: Doubleday.

Conboy, K., Medina, M. and Stanbury, S. (eds) (1997), *Writing on the Body: Female Embodiment and Feminist Theory*. New York: Columbia University Press.

Connolly, W. E (2002), *Neuropolitics: Thinking, Culture, Speed*. Minneapolis: University of Minnesota Press.

Cornell, D. (ed.) (2000), *Feminism and Pornography*. Oxford: Oxford University Press.

Creed, B. (1993), *The Monstrous Feminine: Film, Feminism, Psychoanalysis*. London and New York: Routledge.

Crossley, N. (2001), *The Social Body: Habit, Identity and Desire*. London: Sage.

Crossley, N. (2007), 'Researching Embodiment by Way of "Body Techniques"' in C. Shilling (ed.), *Embodying Sociology: Retrospect, Progress and Prospects*. Oxford: Blackwell Publishing.

Csordas, T. (ed.) (1994), *Embodiment and Experience: The Existential Ground of Culture and Self*. Cambridge: Cambridge University Press.

Darwin, C. (1859), *On the Origin of Species by Means of Natural Selection, or the Preservation of Favoured Races in the Struggle for Life*. 1st edn. London: John Murray.

Darwin, C. (1872), *The Expression of the Emotions in Man and Animals*. London: John Murray.

Davis, K. (1997), *Embodied Practices: Feminist Perspectives on the Body*. London: Sage.

de Beauvoir, S. (1953), *The Second Sex*. New York: Random House.

Deleuze, G. and Guattari, F. (1987), *A Thousand Plateaus: Capitalism and Schizophrenia*. Minneapolis: University of Minnesota Press.

Deleuze, G. (1990), *Negotiations. 1972–1990*. New York: Columbia University Press.

Despret, V. (2004a), *Our Emotional Make-Up: Ethnopsychology and Selfhood*. New York: Other Press.

Despret, V. (2004b), 'The Body We Care for: Figures of Anthropo-zoo-genesis'. *Body and Society*, 10(2–3): 111–34.

Duden, B. (1991), *The Woman Beneath the Skin: A Doctor's Patients in Eighteenth Century German*. Cambridge, MA: Harvard University Press.

Durkheim, E. (1960), 'Le Dualism de la nature humaine et ses conditions sociales' translated in K. H. Wolffe (ed.) *Essays on Sociology and Philosophy*. New York: Harper.

Dyer, R. (1997), *White*. London and New York: Routledge.

Elias, N. (1994), *The Civilizing Process: The History of Manners and State Formation and Civilization*. Oxford: Blackwell.

Elias, N. (2000), *The Civilizing Process: Sociogenetic and Psychogenetic Investigations*. 2nd edn. Oxford: Blackwell.

Falk, P. (1994), *The Consuming Body*. London: Sage.

Fast, J. (1971), *Body Language*. London, Sydney and Auckland: Pan Books.

Featherstone, M., Hepworth, M. and Turner, B. S. (eds) (1991), *The Body: Social Process and Cultural Theory*. London: Sage.

Finnegan, R. (2005), 'Tactile Communication' in C. Classen (ed.), *The Book of Touch*. Oxford: Berg.

Foucault, M. (1973), *The Birth of the Clinic*. London: Tavistock.

Foucault, M. (1977), *Discipline and Punish*. London: Allen Lane.

Franklin, S. (2007), *Dolly Mixtures: The Remaking of Genealogy*. Durham, NC and London: Duke University Press.

Franklin, S., Lury, C. and Stacey, J. (2000), *Global Nature, Global Culture*. London: Sage.

Fraser, M. (2001), 'The Nature of Prozac'. *History of the Human Sciences*, 14(3): 56–84.

Fraser, M. (2003), 'Material Theory. Duration and the Serotonin Hypothesis of Depression'. *Theory, Culture and Society*, 20(5): 1–26.

Fraser, M., Kember, S. and Lury, C. (eds), (2005), 'Inventive Life: Approaches to the New Vitalism, Special Issue. *Theory, Culture and Society*, 22(1).

Fuss, D. (1990), *Essentially Speaking: Feminism, Nature and Difference*. London and New York: Routledge.

Game, A. (2001), 'Riding: Embodying the Centaur'. *Body and Society*, 70(4): 1–12.

Gatens, M. (1996), *Imaginary Bodies: Ethics, Power and Corporeality*. London and New York: Routledge.

Gibbs, A. (2002), 'Disaffected'. *Continnum: Journal of Media and Cultural Studies*, 16(3): 335–40.

Gibson, J. J. (1979), *The Ecological Approach to Visual Perception*. Mahwal, NJ: Lawrence Erlbaum Associates, Inc.

Giddens, A. (1991), *Modernity and Self-Identity: Self and Society in the Late Modern Age*. Cambridge: Polity Press.

Gill, R. (2006), *Gender and the Media*. Cambridge: Polity Press.

Goffman, E. (1959), *The Presentation of Self in Everyday Life*. London: Doubleday.

Greco, M. (1998), *Illness as a Work of Thought*. London and New York: Routledge.

Grosz, E. (1994), *Volatile Bodies: Towards a Corporeal Feminism*. Bloomington, IN: Indiana University Press.

Guattari, F. (1984), *Molecular Revolution: Psychiatry and Politics*. Harmondsworth: Penguin.

Halberstam, J. (1998), *Female Masculinity*. London and Durham, NC: Duke University Press.

Halberstam, J. (2005), *In a Queer Time and Place: Transgender Bodies, Subcultural Lives*. New York: New York University Press.

Haraway, D. (1991), *Simians, Cyborgs and Women: The Reinvention of Nature*. London and New York: Routledge.

Haraway, D. (1996), *Modest_Witness@Second_Millennium.FemaleMan©_Meets_Oncomouse™*. London and New York: Routledge.

Haraway, D. (2003), *The Companion Species Manifesto: Dogs, People and Significant Otherness*. Indiana: Indiana University Press.

Haraway, D. (2004), *The Haraway Reader*. London and New York: Routledge.

Harbord, J. (2002), *Film Cultures*. London and New York: Routledge.

Harbord, J. (2007), *The Evolution of Film: Rethinking Film Studies*. Cambridge: Polity.

Hardt, M. (2007), 'Foreword: What Affects are Good For' in P. Clough with J. Halley (eds), *The Affective Turn: Theorizing the Social*. Durham, NC and London: Duke University Press.

Hardt, M. and Negri, T. (2000), *Empire*. Cambridge, MA: Harvard University Press.

Harre, R. (ed.) (1986), *The Social Construction of Emotion*. New York: Basil Blackwell.

Hatfield, E., Cacioppo, J. T. and Rapson, R. L. (1984), *Emotional Contagion*. Cambridge, New York, Paris: Cambridge University Press.

Hayles, K. (1999), *How We Became Posthuman: Virtual Bodies in Cybernetics and Informatics*. Chicago: Chicago University Press.

Henriques, J., Venn, C., Urwin, C. and Walkerdine, V. (1984), *Changing the Subject: Psychology, Power and Social Regulation*. London: Methuen.

Henriques, J., Venn, C., Urwin, C. and Walkerdine, V. (1998), *Changing the Subject: Psychology, Power and Social Regulation*. 2nd edn. London: Routledge.

Hochschild. A. (1983), *The Commercialization of Intimate Life: Notes from Home and Work*. Stanford, CA: University of California Press.

Hochschild. A. (1994), 'The Commercial Spirit of Intimate Life and the Abduction of Feminism: Signs from Women's Advice Books'. *Theory, Culture and Society*, 11: 1–24.

Hook, D. (2007), *Foucault, Psychology and the Analytics of Power*. Basingstoke and New York: Palgrave.

Howes, D. (2005), 'Skinscapes: Embodiment, Culture and Environment' in C. Classen (ed.), *The Book of Touch*. Oxford: Berg.

Howson, A. (2005), *Embodying Gender*. London: Sage.

Hwa Yol Yung (1996), 'Phenomenology and Body Politics'. *Body and Society*, 2(2): 1–22.

James, W. (1902), *Varieties of Religious Experience: Studies in Human Nature*. London: Routledge.

Johnstone, A. A. (1992) 'The Bodily Nature of the Self or What Descartes Should have Conceded. Princess Elizabeth of Bohemia' in M. Sheets-Johnston (ed.), *Giving the Body its Due*. Albany, NY: State University of New York Press.

Kember, S. (2003), *Cyberfeminism and Artificial Life*. London: Routledge.

Kinsman, G. (1996), 'Responsibility as a Strategy of Governance: Regulating People Living with Aids and Lesbians and Gay Men in Ontario'. *Economy and Society*, 25(3): 393–409.

Kirby, V. (1997), *Telling Flesh: The Substance of the Corporeal*. London and New York: Routledge.

Kraepelin, E. (1913), *Clinical Psychiatry: Lectures and Clinical Psychiatry*. London: Baillière, Tindall and Cox.

Kraepelin, E. (1919), *Dementia Praecox and Paraphrenia*. Edinburgh: E. and S. Livingstone.

Kramer, P. D. (1994), *Listening to Prozac: The Landmark Book about Anti-Depressants and the Remaking of the Self*. London and New York: Penguin Books.

Kristeva, J. (1982), *Powers of Horror: Essay on Abjection*. New York: Columbia University Press.

Kristeva, J (1989), 'Gesture: Practice or Communication?' in T. Polhemus (ed.), *Social Aspects of the Human Body*. Middlesex, New York, Victoria: Penguin Books.

Laing, R.D. (1985), *Wisdom, Madness and Folly. The Making of a Psychiatrist 1927–57*. Edinburgh: Canongate Classics.

Lakoff, A. (2007), 'The Right Patient for the Right Drug: Managing the Placebo Effect in Anti-Depressant Drug Trials'. *Biosocieties*, 2: 57–71.

Laporte, D. (2000), *History of Shit*. Cambridge, MA and London: The MIT Press.

Lash, S. (1991), 'Foucault/Deleuze/Nietzsche' in M. Featherstone, M. Hepworth and B. S. Turner (eds), *The Body: Social Process and Cultural Theory*. London: Sage.

Lash, S. (2006), 'Life (Vitalism)'. *Theory, Culture and Society*, 23(2–3): 323–49.

Latour, B. (2004), 'How to Talk About the Body? The Normative Dimensions of Science Studies'. *Body and Society*, 10(2–3): 205–30.

Latour, B. (2005a), *Reassembling the Social: An Introduction to Actor Network Theory*. Oxford and New York: Oxford University Press.

Latour, B. (2005b), 'From Realpolitik to Drugpolitik or How to Make Things Public' in B. Latour and P. Weibel (eds), *Making Things Public: Atmospheres of Democracy*. Cambridge, MA and London: The MIT Press.

Law, J. (2004), *After Method: Mess in Social Science Research*. London and New York: Routledge.

Law, J. and Mol, A. (eds) (2002), *Complexities: Social Studies of Knowledge Practices*. Durham, NC and New York: Duke University Press.

Lazzarato, M. (2004), 'From Capital-Labour to Capital-Life'. *Ephemera: Theory of the Multitude*, 4(3): 187–208.

Lee, N. and Brown, S. (2002), 'The Disposal of Fear: Childhood, Trauma and Complexity' in J. Law and A. Mol (eds), *Complexities: Social Studies of Knowledge Practices*. Durham, NC and New York: Duke University Press.

Lewis, I. M. (1971), *Ecstatic Religion: A Study of Shamanism and Spirit Possession*. London and New York: Routledge.

Littlewood, R. (1996), 'Reason and Necessity in the Specification of the Multiple Self'. Occasional Paper No. 43. Royal Anthropological Institute of Great Britain and Ireland.

Lupton, D. (1997), 'Foucault and the Medicalisation Critique' in A. Petersen and R. Bunton (eds). *Foucault, Health and Medicine*. London and New York: Routledge.

Lury, C. (1998), *Prosthetic Culture: Photography, Memory, Identity*. London and New York: Routledge.

McNay, L. (1992), *Foucault and Feminism*. Cambridge: Polity Press.

McNeill, W. H. (1995), *Keeping Together in Time: Dance and Drill in Human History*. Cambridge, MA: Harvard University Press.

McRobbie, A. (2000), *Feminism and Youth Culture*. 2nd edn. Basingstoke: Macmillan.

McRobbie, A. (2005), *The Uses of Cultural Studies*. London: Sage.

Manning, E. (2007), *Politics of Touch: Sense, Movement, Sovereignty*. Minneapolis: Minneapolis University Press.

Martin, E. (1987), *The Woman in the Body: A Cultural Analysis of Reproduction*. Boston: Beacon Press.

Martin, E. (2004), 'Talking Back to Neuro-reductionism' in H. Thomas and J. Ahmed (eds), *Cultural Bodies: Ethnography and Theory*. London: Sage.

Massumi, B. (2002a), *Parables for the Virtual: Movements, Affect, Sensation*. Durham, NC and London: Duke University Press.

Massumi, B. (ed.) (2002b), *A Shock to Thought: Expression after Deleuze and Guattari*. London and New York: Routledge.

Mead, G. H. (1934), *Mind, Self and Society*. Chicago: Chicago University Press.

Merleau-Ponty, M (1962), *Phenomenology of Perception*. London: Routledge and Kegan Paul.

Metzl, J. M. (2003), *Prozac on the Couch: Prescribing Gender in the Era of Wonder Drugs*. Durham, NC and London: Duke University Press.

Miller, W. I (1997), *The Anatomy of Disgust*. Cambridge, MA and London: Harvard University Press.

Moerman, D. E. (1992), 'Minding the Body: The Placebo Effect Unmasked' in M. Sheets-Johnston (ed.). *Giving the Body its Due*. Albany, NY: State University of New York Press.

Mol, A. (2002), *The Body Multiple: Ontology in Medical Practice*. London and New York: Duke University Press.

Mol, A. and Law, J. (2004), 'Embodied Action, Enacted Bodies: The Example of Hypoglycaemia'. *Body and Society*, 10(2–3): 43–62.

Nettleton, S. (1992), *Power, Pain and Dentistry*. Buckingham and Philadelphia: Open University Press.

Nottingham, S. (2000), *Screening DNA: Exploiting the Cinema–Genetics Interface*. http://www.dnabooks.co.uk: DNA Books.

Oakley, A. (1984), *The Captured Womb: A History of the Medical Care of Pregnant Women*. Oxford: Basil Blackwell.

Orr, J. (2006), *Panic Diaries: A Genealogy of Panic Disorder*. New York: Duke University Press.

Petersen, A. and Bunton, R. (eds) (1997), *Foucault, Health and Medicine*. London and New York: Routledge.

Pfungst, O. (2000), *Clever Hans (The Horse of Mr Von Osten)*. Bristol: Thoemmes Press.

Porter, R. (2003), *Flesh in the Age of Reason*. London, New York, Victoria, Ontario, New Delhi, Auckland and Rosebank, SA: Penguin, Allen Lane.

Potter, J. and Wetherell, M. (1987), *Discourse and Social Psychology: Beyond Attitudes and Behaviour*. London: Sage.

Probyn, E. (2005), *Blush: Faces of Shame*. Minneapolis: University of Minnesota Press.

Prosser, J. (1998), *Second Skins: The Body Narratives of Transsexuality*. New York: Columbia University Press.

Rabinow, P. (ed.) (1991), *The Foucault Reader: An Introduction to Foucault's Thought*. London: Penguin.

Riefenstahl, L. (1934), *The Triumph of the Will*. Produced and Directed by Leni Riefenstahl.

Riley, D. (1983), *War in the Nursery*. London: Virago Press.

Romme, M. and Escher, S. (eds) (1993), *Accepting Voices*. London: Mind.

Rose, H. (2007), 'Eugenics and Genetics: The Conjoint Twins'. *New Formations*, 60: 13–26.

Rose, N. (1989), *Governing the Soul: The Shaping of the Private Self*. London: Routledge.

Rose, N. (1996), *Inventing Ourselves: Psychology, Power and Personhood*. Cambridge: Cambridge University Press.

Rose, N. (1999), *Powers of Freedom: Reframing Political Thought*. Cambridge: Cambridge University Press.

Rose, N. (2004), 'Becoming Neurochemical Selves' in Nico Stehr (ed.), *Biotechnology: Between, Commerce and Civil Society*. New York: Transaction Press.

Rose, N. (2007), *The Politics of Life Itself: Biomedicine, Power and Subjectivity in the 21st Century*. Princeton: Princeton University Press.

Rosenthal, R. (1966), *Experimenter Effects in Behavioural Research*. New York: Appleton Century Crofts.

Sargeant, W. (1967), *The Unquiet Mind*. London: Pan Books.

Sedgwick, E. K. (2003), *Touching Feeling: Affect, Pedagogy, Performativity*. Durham, NC and London: Duke University Press.

Shapira, K. H. A., McClelland, N. R., Griffiths and Newell, J. D. (1970), 'A Study of Effects of Tablet Colour in Treatment of Anxiety States'. *British Medical Journal*, 2: 446.

Sheets-Johnston, M. (ed.) (1992), *Giving the Body its Due*. Albany, NY: State University of New York Press.

Sheets-Johnston, M. (1999), *The Primacy of Movement*. Amsterdam and Philadelphia: John Benjamin's Publishing Co.

Sheldrake, R. (2000), *Dogs That Know When Their Owners are Coming Home, and Other Unexplained Powers of Animals*. New York: Three Rivers Press.

Schiebinger, L. (ed.) (2000), *Feminism and the Body*. Oxford: Oxford University Press.

Shilling, C. (1993), *The Body and Social Theory*. London, Thousand Oaks, CA and New Delhi: Sage.

Shilling, C. (2003), *The Body and Social Theory*. 2nd edn. London, Thousand Oaks, CA and New Delhi: Sage.

Simondon, G. (1992), 'The Genesis of the Individual' in J. Crary and S. Kwinter (eds), *Incorporations*. New York: Zone Books.

6, P., Radstone, S., Squire, C. and Treacher, A. (2007), *Public Emotions*. Basingstoke and New York: Palgrave Macmillan.

Skeggs, B. (1997), *Formations of Class and Gender: Becoming Respectable*. London, Thousand Oaks, CA and New Delhi: Sage.

Skeggs, B. (2004), *Class, Self and Culture*. London and New York: Routledge.

Smith, R. (1992), *Inhibition: History and Meaning in the Sciences of Mind and Brain*. Berkeley: University of California Press.

Sobchack, V. (2004), *Carnal Thoughts: Embodiment and Moving Image Culture*. Berkeley: University of California Press.

Spence, J. (1995), *Cultural Sniping: The Art of Transgression*. London and New York: Routledge.

Stacey, J. (1997), *Teratologies: A Cultural Study of Cancer*. London and New York: Routledge.

Stafford, B. (2007), 'Self-Eugenics: The Creeping Illusionising of Identity from Neurobiology to Newgenics'. *New Formations*, 60: 102–11.

Stenner, P. (1993), 'Discoursing Jealousy' in E. Burman and I. Parker (eds), *Discourse Analytic Research: Repertoires and Readings of Texts in Action*. London: Routledge.

Tamborinino, J. (2002), *The Corporeal Turn: Passion, Necessity, Politics*. Lanham, Boulder, New York and Oxford: Rowman and Littlefield Publishers Inc.

Thacker, E. (2003), 'What is Biomedia?' *Configurations*, 11: 47–79.

Thrift, N. (2000), 'Still Life in Nearly Present Time: The Object of Nature'. *Body and Society*, 6: 34–57.

Thrift, N. (2004), 'Intensities of Feeling: Towards a Spatial Politics of Affect'. *Geografiska Annaler*, 86B(1): 55–76.

Thrift, N. (2007), *Non-Representational Theory: Space, Politics, Affect*. London and New York: Routledge.

Tomkins, S. (1962), *Affect, Imagery, Consciousness*. Vol. 1. New York: Springer.

Turner, B. (1984), *The Body and Society: Explorations in Social Theory*. 1st edn. Oxford and New York: Blackwell.

Turner, B. (1991), 'Recent Developments in the Theory of the Body' in M. Featherstone, M. Hepworth and B. S. Turner (eds), *The Body: Social Process and Cultural Theory*. London, Newbury Park, New Delhi: Sage.

Turner, B. (1996), *The Body and Society: Explorations in Social Theory*. 2nd edn. London, Thousand Oaks, CA: Sage.

Walkerdine, V. (1990), *Schoolgirl Fictions*. London: Verso Press.

Walkerdine, V. (1997), *Daddy's Girl: Young Girls and Popular Culture*. Basingstoke: Macmillan.

Walkerdine, V. (2007), *Children, Gender, Video Games*. Basingstoke and New York: Palgrave Macmillan.

Watkins, E. S. (2007), 'Passing the Postmenopausal Pregnancy: A Case Study in the New Eugenics'. *New Formations*, 60: 102–11.

Williams, S. J. (2005), *Sleep and Society: Sociological Ventures into the (Un)known*. Didcot: Routledge.

Williams, S. J. and Bendelow, G. (1998), *The Lived Body: Sociological Themes, Embodied Issues*. London and New York: Routledge.

Wilson, E. A. (2004), *Psychosomatic: Feminism and the Neurological Body*. Durham, NC and London: Duke University Press.

Wissinger, E. (2007), 'Modelling a Way of Life: Immaterial and Affective Labour in the Fashion Modelling Industry'. *Ephemera*, 7(1): 250–69.

Woodward, K. (ed.) (1997), *Identity and Difference*. London, Thousand Oaks, CA and New Delhi: Sage.

Young, I. (1990), *Throwing Like a Girl and Other Essays in Feminist Philosophy and Social Theory*. Bloomington, IN: Indiana University Press.

Zylinska, J. (2005), *The Ethics of Cultural Studies*. London and New York: Continuum.

INDEX

abjection, 12, 92–4, 100
actor network theory, 120–1, 124
affect, 10, 13, 52–3, 57, 65, 133
    affective body, 9–10, 132
        see also bodily affectivity
    affective comportment, 67, 69–70
    affective exchange, 65
    affective self-containment, 44
    affective transmission, 49
affective labour, 134
agency, 16–17, 28, 131
alterity, 117, 120
    see also otherness
Anzieu, D., 86–7, 110
arteriosclerosis, 125–6
articulation, 97, 105–6, 108, 110, 121
    articulated body, 96
assemblage(s), 97, 110, 122–4, 133
atavism, 19
attunement, 9, 40–1, 53, 87, 119, 132

Beck, U., 70
becoming, 13, 38, 40–1, 44, 46, 50, 53, 78–9,
        94, 105, 110, 122
    becoming-other, 111
    see also feminine becoming
becoming woman, *see* feminine becoming
Bendelow, G., 85, 93
Big Brother, 45
Biologism, 18–19
biology, *see* socialized biology
biomediation, 136–7
bio-power, 75
Blackman, L., 53, 60, 68, 127–9, 133, 135
bodies without organs, 109–12
bodily affectivity, 64–5, 67–8, 70, 74, 76, 78

body as
    absent-presence, 6–7, 13, 15
    articulated, *see* articulated body
    civilized, 50–1
    difference, 11
    molecular, *see* molecular body
    normative, 12
    multiple, 13
        see also multiplicity
    process, 80, 107
        see also process
    sensient, 84
body assemblages, *see* assemblage(s)
body image, 77
body language, 41–2
    body leakage, 42–3
    body tells, 43
    see also non-verbal communication
body practices, 75, 108
    see also body techniques
body studies, 3, 7, 16, 20
body techniques, 108
    see also body practices
body theory, 7, 20
bodyself, 23
Bordo, S., 29, 75
Bourdieu, P., 62–3, 88
Braidotti, R., 74, 76, 78–80, 94, 109–11
Brennan, T., 44, 47–9, 51–3
Buñuel, L., 90
Butler, J., 78–80

Cartesian dualism, *see* dualism
Charlesworth, S., 64–5
cloning, 112–13, 116
Clough, P., 133, 135

companion species, 118–20
contagious communication, 137
conversation analysis, 23
corporeality, 7, 20, 29–30, 49–50, 53, 56, 105, 118
   corporeal consciousness, 83
   corporeal turn, 72
   corporeal feminism, 73–4, 76, 78
cosmetic surgery, 27
critical psychology, 5, 23, 27, 48–9
Crossley, N., 63, 108
cultural influence, 5
cultural inscription, 16–17, 20, 22, 26, 32, 49, 60, 71
   see also social constructionism

dance, 107–9
   see also tango
Darwinistic evolutionism, 17–18, 45, 48
de Beauvoir, S., 73
degeneracy, 18–19
Deleuze, G., 110
Deleuzian philosophy, 78, 80, 109
depression, 114
desire, 68
Despret, V., 11, 40–1, 44, 46, 50–1, 53
de-territorilization, 111
disciplinary power, 25, 28–9, 39
discourse, 24
   discourse-analysis, 23
   discourse determinism, 28, 61
disgust, 94
docile body, 26–7, 30, 75, 134
dualism, 4, 6–8, 17, 21, 27
   Cartesian, 10, 21, 55, 57, 75, 84
   mind-body, 4, 10, 22, 24, 32, 52, 70, 75, 84, 92, 126
Durkheim, E., 16–17, 20

eating disorders, 26, 29, 38–9
Elias, N., 51, 57
embodied subjectivity, see subjectivity
embodiment, 12, 34, 37, 49, 57, 103, 121
emotion management, 42
emotional contagion, 46–7, 137
enactment, 10, 12–13, 105, 129
   enacted materialities, 129

essentialism, 19–21, 28–9, 31
ethnomethodology, 23
eugenics, 18
experience, 83, 89

Falk, P., 84–5, 88–91
Featherstone, M., 17
feeling, 133
feminine becoming, 72, 74–7, 111
feminism, 27, 72
   see also corporeal feminism
fiction of the autonomous self, 70, 76, 91, 113
   see also individualization
figuration, 109
Foucault, M., 24–7, 29, 33, 134
   discipline and punish, 26
functionalism, 16

Game, A., 8–9, 39–41, 46, 119, 132
Giddens, A., 70
Goffman, E., 42–4
   impression management, 43
Grosz, E., 76–8
Guattari, F., 100
gut feeling, 53

Hans the Horse, see horse–human relations
Haraway, D., 109, 111, 119–20
Harry Potter, 66, 70
Hayles, K., 117–18
healthism, 12, 98–9, 101
   see also self-health
hearing voices network, 127–8
   see also psychiatric culture
HIV and AIDS, 11
Hochschild, A., 42
Horse–human relations 8, 38–40
   Hans the Horse, 11, 38, 44, 46, 119, 132
hypoglycaemia, 123

immateriality, 132, 134
imposition, 16, 25
inculcation, 25
individualization, 70, 92
   individualized body, 94
   individualized self, 9
internalization, 74

James, W., 106

kinaesthetic body, 64
  kinaesthesia, 83
Kristeva, J., 93

Laing, R.D., 9
Laporte, D., 95
Latour, B., 96–7, 106, 121–2
Law, J., 123–4, 126
lived body, 12, 83, 89, 92, 98, 121
lived experience, 12

Manning, E., 108, 110, 119
Massumi, B., 109–10
materiality, 7, 20, 29, 34, 50, 56, 72, 105,
  118, 135
McRobbie, A., 60
Mead, G.H., 22
  looking-glass self, 23
Merleau Ponty, M., 64
mind-body dualism, see dualism
mind-over-matter, 5–6, 9, 55–6
modulation, 125
Mol, A., 123–6
molecular body, 113–14, 118
mouth, 89
movement, 105–6, 108, 110
multiplicity, 105–6, 109, 124
muscular bonding, 30–1, 85, 134

naturalistic body, 2, 17, 20–1, 37, 45, 85, 88
naturecultures, 120
networked body, 56
non-verbal communication, 41–2, 45
  see also body language
otherness, 59–60, 64, 70, 89, 110
  see also alterity

Panopticon, 25
performance, 13
performativity, 74, 79
phenomenology, 64, 66
placebo effect, 54–5
post-colonialism, 27
posthuman bodies, 116–18
power, 24, 131
  as repression, 25

practice(s), 5, 13, 123–5, 128–9
process, 5, 12, 105–6
  see also body as process
prosthetic selfhood, 117
psychiatric culture, 127
  see also hearing voices network
psychosomatic, 55–6

relationality, 34–5, 40, 51, 55–6, 77, 120
rhythm, 9
risk society, 11
Rose, N., 70, 90–1, 107, 113–15, 118, 128

self, 22
self-health, 98–9, 101
  see also healthism
Sheets-Johnstone, M., 53–5, 106–7
Shilling, C., 17, 20–1, 27–9, 38, 43–4, 51
Simondon, G., 109
Skeggs, B., 59, 61–3
skin ego, 86–7, 110
skin knowledge, 86–7, 108
sleep, 32–3, 43
smell, 95
social constructionism 6, 21–4, 27–9, 49, 71,
  77
  see also cultural inscription
social influence, 11, 37–40, 44, 46, 50, 122
socialized biology, 126–7
socio-biology, 21
sociology of the body, 2–3, 10, 51, 57
somatic, 7, 20, 105, 118
  somatic individuality, 12, 114, 128
  somatically felt body, 10, 20, 29–30, 52,
    54
Spence, J., 98
Stacey, J., 12, 99, 102
subjectification, 24
subjectivity, 3, 28, 48, 69, 72, 80, 89, 131,
  138
  embodied subjectivity, 74, 118
symbolic violence, 60
synesthesia, 84

tango, 107, 109, 119
taste, 88
territorialization, 111
Tomkins, S., 136–7

touch, 108
transdisciplinary, 7–8, 53
Turner, B., 15–17, 20, 22–4, 27, 32, 35,
  43–4
twinning, 112–13, 116

vitalist body, 54–5, 57

Walkerdine, V., 60, 68–9
will, 4, 9, 22, 32, 43, 57, 70, 75
Williams, S., 32–4, 35, 85, 93